Fundamentals of Electronics 1

Fundamentals of Electronics 1

Electronic Components and Elementary Functions

Pierre Muret

iSTE

WILEY

First published 2017 in Great Britain and the United States by ISTE Ltd and John Wiley & Sons, Inc.

ISTE Ltd
27-37 St George's Road
London SW19 4EU
UK

www.iste.co.uk

John Wiley & Sons, Inc.
111 River Street
Hoboken, NJ 07030
USA

www.wiley.com

Library of Congress Control Number: 2017944816

British Library Cataloguing-in-Publication Data
A CIP record for this book is available from the British Library
ISBN 978-1-78630-181-9

Contents

Preface

Today, we can consider electronics to be a subject derived from both the theoretical advances achieved during the 20th Century in areas comprising the modeling and conception of components, circuits, signals and systems, together with the tremendous development attained in integrated circuit technology. However, such development led to something of a knowledge diaspora that this work will attempt to contravene by collecting both the general principles at the center of all electronic systems and components, together with the synthesis and analysis methods required to describe and understand these components and subcomponents. The work is divided into three volumes. Each volume follows one guiding principle from which various concepts flow. Accordingly, Volume 1 addresses the physics of semiconductor components and the consequences thereof, that is, the relations between component properties and electrical models. Volume 2 addresses continuous time systems, initially adopting a general approach in Chapter 1, followed by a review of the highly involved subject of quadripoles in Chapter 2. Volume 3 is devoted to discrete-time and/or quantized level systems. The former, also known as sampled systems, which can either be analog or digital, are studied in Chapter 1, while the latter, conversion systems, we address in Chapter 2. The chapter headings are indicated in the following general outline.

Each chapter is paired with exercises and detailed corrections, with two objectives. First, these exercises help illustrate the general principles addressed in the course, proposing new application layouts and showing how theory can be implemented to assess their properties. Second, the exercises act as extensions of the course, illustrating circuits that may have been described briefly, but whose properties have not been studied in detail. The first volume should be accessible to students with a scientific literacy corresponding to the first 2 years of university education, allowing them to acquire the level of understanding required for the third year of their electronics degree. The level of comprehension required for the following two volumes is that of students on a master's degree program or enrolled in engineering school.

In summary, electronics, as presented in this book, is an engineering science that concerns the modeling of components and systems from their physical properties to their established function, allowing for the transformation of electrical signals and information processing. Here, the various items are summarized along with their properties to help readers follow the broader direction of their organization and thereby avoid fragmentation and overlap. The representation of signals is treated in a balanced manner, which means that the spectral aspect is given its proper place, to do otherwise would have been outmoded and against the grain of modern electronics, since now a wide range of problems are initially addressed according to criteria concerning frequency response, bandwidth and signal spectrum modification. This should by no means overshadow the application of electrokinetic laws, which remains a necessary first step since electronics remains fundamentally concerned with electric circuits. Concepts related to radio-frequency circuits are not given special treatment here, but can be found in several chapters. Since the summary of logical circuits involves digital electronics and industrial computing, the part treated here is limited to logical functions that may be useful in binary numbers computing and elementary sequencing. The author hopes that this work contributes to a broad foundation for the analysis, modeling and synthesis of most active and passive circuits in electronics, giving readers a good start to begin the development and simulation of integrated circuits.

Outline

1) Volume 1: Electronic components and elementary functions.

   i) Diodes and applications

   ii) Bipolar transistors and applications

   iii) Field effect transistor and applications

   iv) Amplifiers, comparators and other analog circuits

2) Volume 2: Continuous-time signals and systems [MUR 17a].

i) Continuous-time stationary systems: General properties, feedback, stability, oscillators

ii) Continuous-time linear and stationary systems: Two-port networks, filtering and analog filter synthesis

3) Volume 3: Discrete-time signals and systems and conversion systems [MUR 17b].

i) Discrete-time signals: Sampling, filtering and phase control, frequency control circuits

ii) Quantized level systems: Digital-to-analog and analog-to-digital conversions

Pierre MURET
June 2017

Introduction

In this first volume, we address the physics of semiconductor devices directly through electrostatics and the laws steering the transport of charge carriers. This allows us to broach the governing principles at work in semiconductor electric components, which are described and quantified in order to expose the relations that control external electric quantities. Through the first three chapters, devoted, respectively, to diodes, bipolar transistors and field effect transistors, the electrical characteristics of these components can be deduced and expressed as electric models. We then consider the static or instantaneous voltages and currents present in these nonlinear or linear electric models to obtain the properties of active circuits made using these components and passive elements. In some cases, these considerations will require students to solve first-order differential equations or use complex impedance and transmittance to represent values in sinusoidal conditions (see Appendix).

Each chapter emphases the applications derived from these models, in terms of analog and logical functions, a majority of which are based on the original nonlinearity of the components. Conversely, linearized models become useful mainly from Chapter 2 onwards, to help with in-depth analysis of the amplifier circuits, and in particular for the operational amplifiers in Chapter 4. In this volume, the study of signal representations is kept to a strict minimum, largely confined to the Appendix, with the exception of the noise problem in amplifiers in Chapter 4. The content of the Appendix should be of use for the wider public who may be unfamiliar with the laws and theorems of electrokinetics, which are indispensable in this volume covering a variety of electronic circuits and equivalent component circuits. Exercises allow a deeper analysis of fundamental layouts comprising a small number of active devices, typically from two to 10.

1

Diodes and Applications

1.1. Semiconductor physics and current transport in pn diodes

1.1.1. *Energy and concentration of mobile charge carriers (electrons and holes)*

Studies of electrons' physical properties indicate that they appear either as particles with a movement quantity (or impulse) of p and mass m, or as waves of wavelength λ and wave vector k. Between these values, De Broglie's wave mechanics establishes the following relation: $p = \hbar k = \dfrac{h}{\lambda}$, where $h = 6.62 \times 10^{-34}$ J·s and $\hbar = h / 2\pi$.

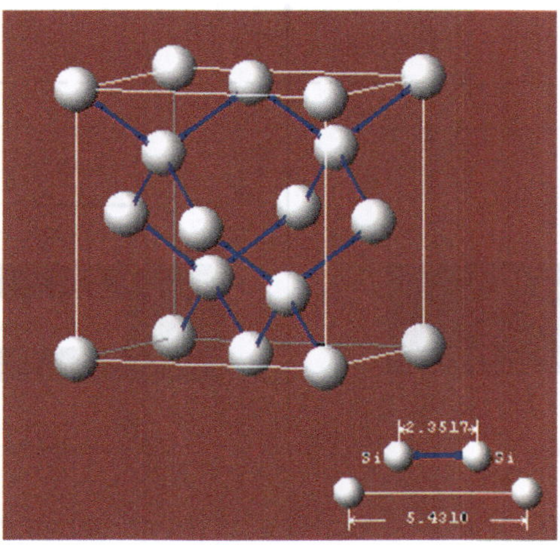

Figure 1.1. *Silicon crystal lattice (Ångstrom distances, equal to 0.1 nm)*

Here, kinetic energy is $E_C = \dfrac{p^2}{2m}$, while potential energy corresponds to the work of the attractive force between 1 electron and 1 proton until they are approximately separated by the atomic radius, that is $a_0 \approx 0.2$ nm; therefore, $U = \int_{a_0}^{\infty} \dfrac{e^2}{4\pi\varepsilon r^2} dr = \dfrac{e^2}{4\pi\varepsilon a_0}$ (with $e = 1.6 \times 10^{-19}$ Coulomb and $\varepsilon \approx 10^{-10}$ Farad/m in a semiconductor such as Si) to the order of 10^{-19} J, or a little less than 1 eV.

In an isolated atom, quantum mechanics makes a connection between energy and the wave frequency associated with each electron, so that the energy of each can only take certain values known as energy levels. When the atoms are in a solid such as silicon, which can hold a crystal shape where atoms are arranged in a regular and periodic manner in space (Figure 1.1), the potential in terms of the electrons is that determined by the atoms' nuclei as well as other electrons, also becoming regular and periodic in space.

A consequence of this is the transformation of energy levels into allowed energy bands separated by a forbidden bandgap (Figure 1.2). These allowed bands are made up of as many energy levels as there are electrons in the solid, also known as energy states or quantum states, spread over an energy range of several electron volts.

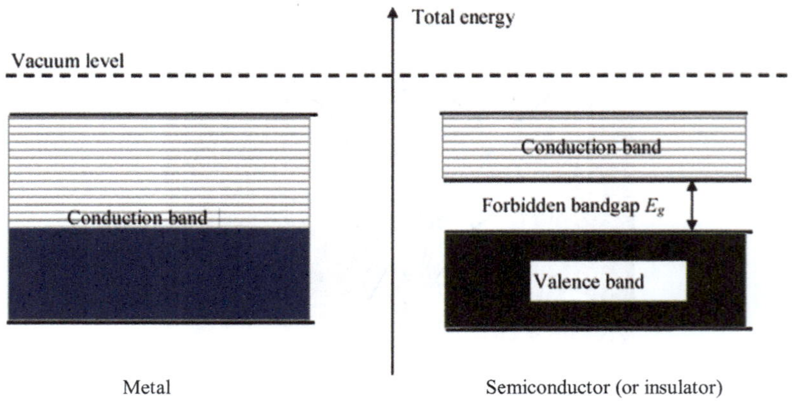

Figure 1.2. *Energy bands of a solid (full and empty quantum states in dark and hatched color, respectively)*

Only a single electron can be placed in the allowed bands per quantum state, both for isolated atoms and in solids.

Conduction is only possible if electrons can change quantum state, as this allows them to acquire kinetic energy and movement. This change can occur in the case of metals, as the allowed band with the highest energy levels, known as the conduction band, is only ever partially filled; conversely, this can only occur when the temperature increases above absolute zero in semiconductors, since the forbidden band separates a full valence band from an empty conduction band at absolute zero. This is because thermal excitation induces the transfer of some electrons from the valence band into the conduction band. In the case of semiconductors, statistics show that the product of electron concentration n in the conduction band and of holes p (that is the absence of electrons that may be considered as positively charged particles with a positive mass) in the valence band is equal to the square of the intrinsic concentration n_i:

$$n.p = n_i^2 = N_c N_v \, exp(-E_g/kT)$$

where E_g = width of the forbidden bandgap, k = Boltzmann constant = 1.38 × 10^{-23} J/K and N_c, N_v = density of effective state in the conduction and valence bands (m^{-3}) so that $N_c N_v = B\,T^3$ (T in °K and B to the order of 5 × 10^{43} m^{-6} K^{-3} for silicon).

Solids with semiconductor characteristics are chiefly those whose atoms have 4 electrons on their peripheral layer, that is those in column IV of the periodic table ($2s^2\,2p^2$ configuration for diamond, $3s^2\,3p^2$ for silicon, $4s^2\,4p^2$ for germanium, or mixed for SiC) or also those made up of atoms from columns III and V (called III–V, such as GaAs, GaP, InP, InAs, GaN, AlN and InN) or columns II and VI (called II–VI, such as ZnO, ZnSe, CdTe, CdS and ZnTe).

Accordingly, we can only consider the peripheral electrons, of which there are on average 4 per atom, whose charge is balanced by four nucleus protons (either 3 and 5, or 2 and 6 for materials III–V and II–VI, respectively).

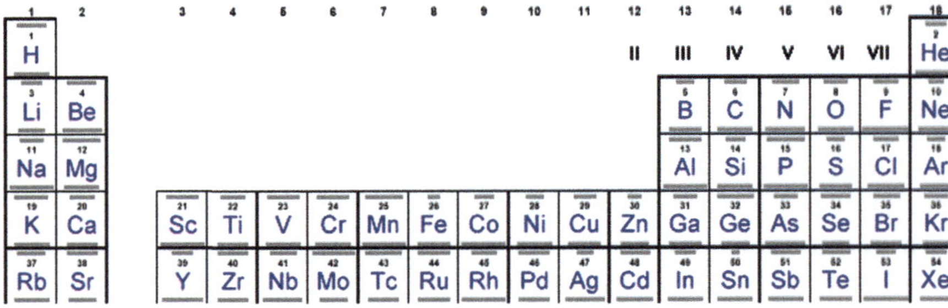

Figure 1.3. *Periodic table of elements*

In intrinsic semiconductors, that is ideally without any impurities, there are as many holes as electrons, such that $n = p = n_i$, the density of intrinsic carriers, since an excited electron in the conduction band automatically leaves a hole in the valence band (Figure 1.4).

This situation is modified by doping, which corresponds to the introduction of foreign atoms, also known as impurities. Doped semiconductors are far more useful, since we can favor either conduction by electrons or conduction by holes. The relation $n{\cdot}p = n_i^2$ still applies, however:

– for n doping of semiconductors IV-IV: $n = N_D >> p$ by introducing atoms located in a column further to the right, which then become a concentration of N_D donors;

– for p doping of semiconductors IV-IV: $p = N_A >> n$ by introducing atoms located in a column further to the left, which then become a concentration of N_A acceptors. For III–V and II–VI, doping occurs along the same lines, that is by increasing or decreasing the number of nucleus protons by one unit with the impurity relative to the atom that is being replaced.

By means of doped semiconductors (Figure 1.5), we can obtain zones with localized charges (ionized donors or acceptors) if the mobile carriers (electrons or holes) have been carried into another zone of the component by an electric field.

The combination of several such zones, known as space-charge zones or more commonly, depleted zones, also allows us to obtain this electric field and create components. Moreover, a p type zone in contact with an n type zone forms a pn diode. The asymmetry of fixed charges in both zones leads to asymmetry of electrical characteristics and the rectification effect.

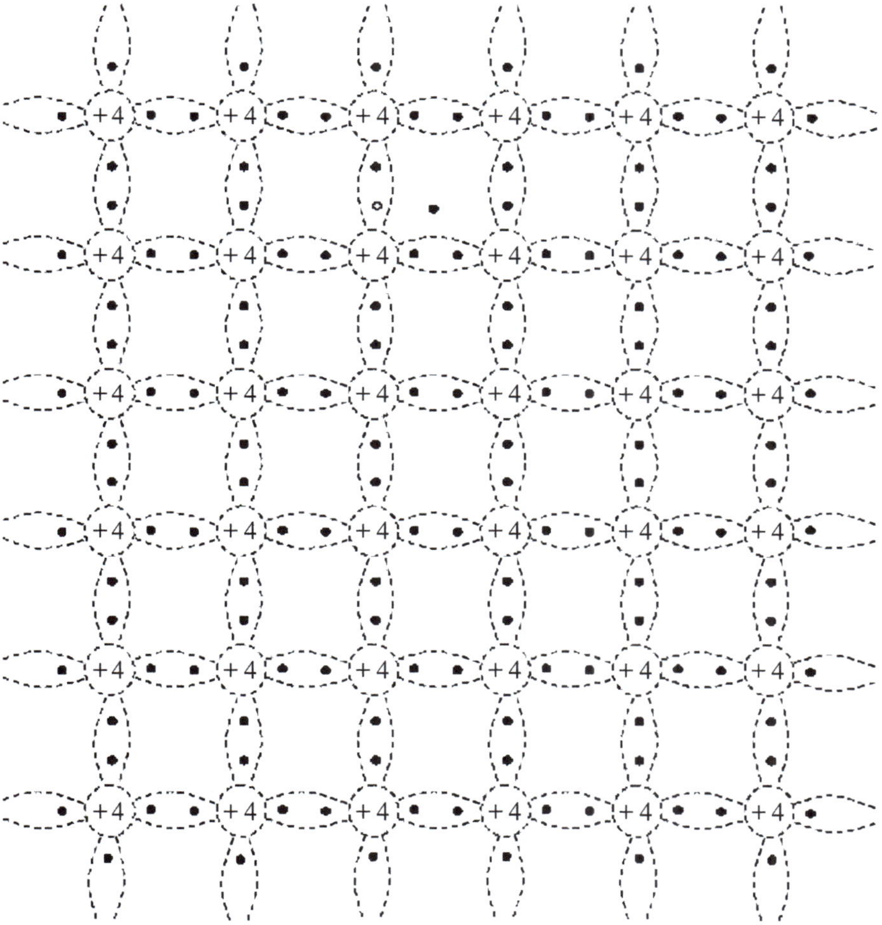

Figure 1.4. *Flattened structure of a doped intrinsic covalent semiconductor (electrons represented by a full black circle, holes by a hollow black circle)*

The fixed charges that remain in a depletion zone after the departure of electrons are positive due to the surplus positive charge (5+ on the proton and 4– for the peripheral electrons) on the nucleus of ionized donors. On the other hand, the fixed charges that remain in a type p depletion zone after the departure of holes are negative due to the electron that has taken the place of the hole when it leaves the ionized acceptor (3+ on the proton and 4– for the peripheral electrons), which is a surplus negative charge (Figure 1.5). In an n-type semiconductor, the electrons are majority carriers while the holes are minority carriers, and the opposite is true for a p-type semiconductor.

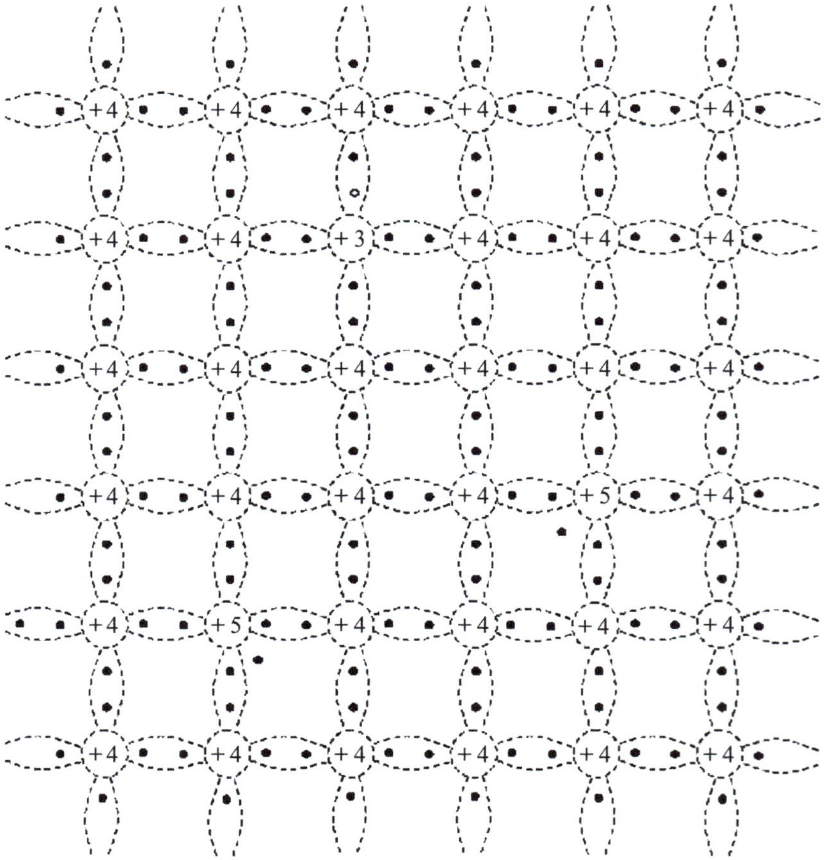

Figure 1.5. *Flattened structure of a doped covalent semiconductor (electrons represented by a full black circle, holes by a hollow black circle)*

1.1.2. Conduction mechanisms

In neutral semiconductors and depletion zones, we see two types current density (in A/m^2 or A/cm^2) in steady state:

– a conduction current density $en\mu_n \mathbf{E}$ for electrons and $ep\mu_p \mathbf{E}$ for holes, or in total $\mathbf{J}_c = e(n\mu_n + p\mu_p)\mathbf{E}$, where μ_n and μ_p is the mobility of electrons and holes, respectively, and **E** is the electric field vector, which gives rise to the Ohm law in conductors.

– a diffusion current density: $eD_n \underline{grad}(n) = \mu_n kT\ \underline{grad}(n)$ for electrons, $-eD_p grad(p) = -\mu_p kT\ \underline{grad}(p)$ for holes, with D_n and D_p representing the diffusion coefficients according to the Einstein relation $D_n = \mu_n kT/e$ and $D_p = \mu_p kT/e$, that is, in total $J_d = \mu_n kT\ \underline{grad}(n) - \mu_p kT\ \underline{grad}(p)$.

However, the spatial shape of potential in the depletion zones will determine these currents, since it is here that we find the strongest electric fields and charge carrier concentration gradients. We can discern the two p and n zones of the pn diode, each of which is divided to one neutral zone in contact with the external electrodes and one depletion zone charged by ionized impurities, positive on the n side and negative on the p side. These zones contain concentrations of the carriers represented in Figure 1.6.

The difference in carrier concentrations is very significant in the neutral zones and in most parts of the depletion zones. In addition, in the absence of any voltage applied to the diode (so at a total current of zero), the total current in terms of conduction and diffusion for each type of carrier must be cancelled out, since the electrons and holes are independent in an ideal diode. As we will see in section 1.4.1, one consequence of this is that current I_d in the diode depends exponentially on the external applied potential V_d. There exists a charge of minority carriers injected into each neutral zone that is proportional to I_d in forward bias ($V_d > 0$), when the number of majority charges is increased, with a diffusion current that is predominant:

$$I_{\text{diff}} = I_0(T)\exp(eV_d/kT)$$

where V_d = voltage applied at the p terminal relative to the n terminal. We write the thermal voltage $V_T = kT/e = 26$ mV at $T = 300$ K in the case of an ideal diode.

There is also a minority conduction current $-I_0(T)$. So, finally we write:

$$I_d = I_0(T)[\exp(V_d/V_T) - 1]$$

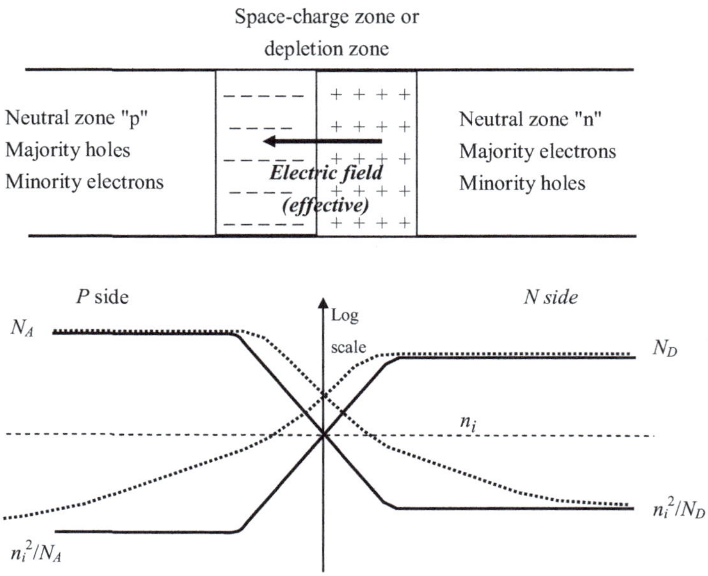

Solid line = diode at equilibrium (zero external voltage difference) ;
Dotted line = forward polarization

Figure 1.6. *Fixed charges and concentration of holes and electrons in the pn diode*

In reverse bias, only the $-I_0(T)$ current remains when $V_d < -3V_T$. The depletion zone can be approximated by an insulator, resulting in a capacitance that varies in accordance with V_d, as detailed below.

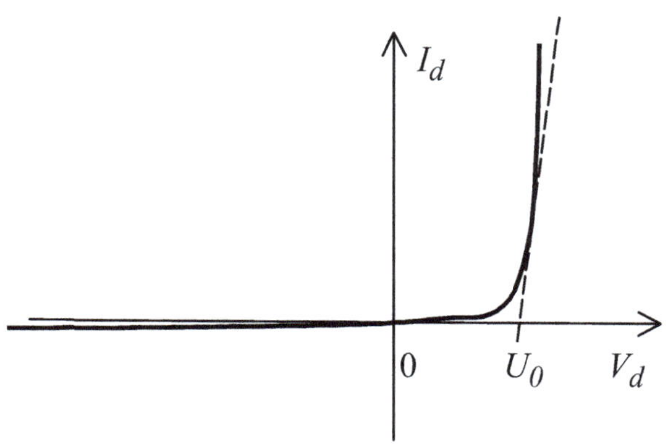

Figure 1.7. *Exponential characteristic of pn diode*

1.2. Simplified models of the pn diode

In forward bias, if we equate the characteristic $I_d(V_d)$ to its tangent at the operating point, the equation of this tangent is $V_d = U_0 + R_d I_d$, where U_0 is the diode threshold (Figure 1.7) and R_d its dynamic resistance equal to the inverse of the slope of $I_d(V_d)$.

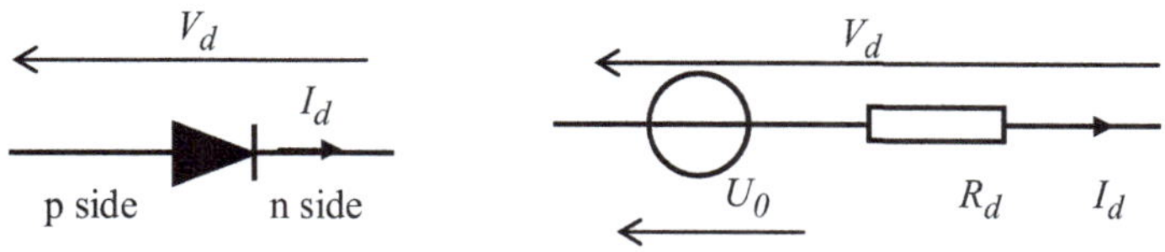

Figure 1.8. *Symbolic and electric representations (in forward bias) of the pn diode*

In certain cases, we can take $R_d = 0$ and even $U_0 = 0$ to simplify the first stage of analysis. However, in reality $R_d^{-1} = \frac{\partial I_d}{\partial V_d} = \frac{I_0}{V_T}\exp\left(\frac{V_d}{V_T}\right) = \frac{I_d}{V_T}$ or alternatively: $R_d = \frac{V_T}{I_d}$.

In reverse bias, we can consider the diode as an open circuit as a first approximation.

Accordingly, the diode is the dipole with the most basic commutation role. *Its commutation is driven by the controlling voltage V_d*, so that, with the simplified model:

– the diode is an open circuit (or off) if $V_d < U_0$ and the diode is a closed circuit (or on) if $V_d > U_0$.

So, in a circuit with (at least) one diode, we must analyze the controlling voltage to determine its state, which can require a starting hypothesis to be verified through analysis.

1.3. Diode circuitry (reviewed in exercises)

Several examples of the electronic operations performed with diodes are provided below. Remember that since diodes are commutation devices controlled by voltage applied to them, either the on or off state of each diode must be subject to a working hypothesis that is confirmed or refuted further according to the consequences of the hypothesis. This step allows us to analysis the following circuits (Figures 1.9 and 1.10).

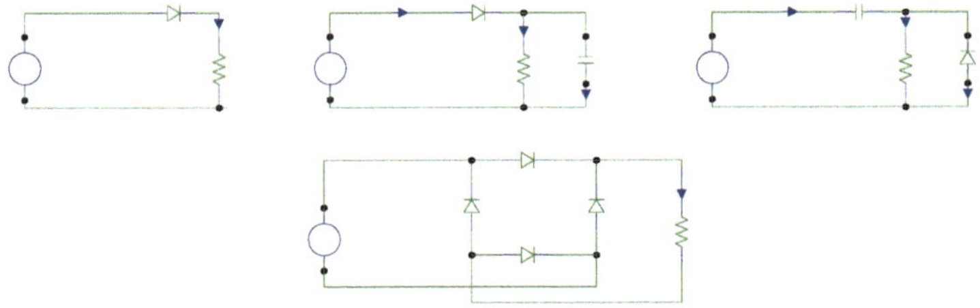

Figure 1.9. *Rectification and detection operations*

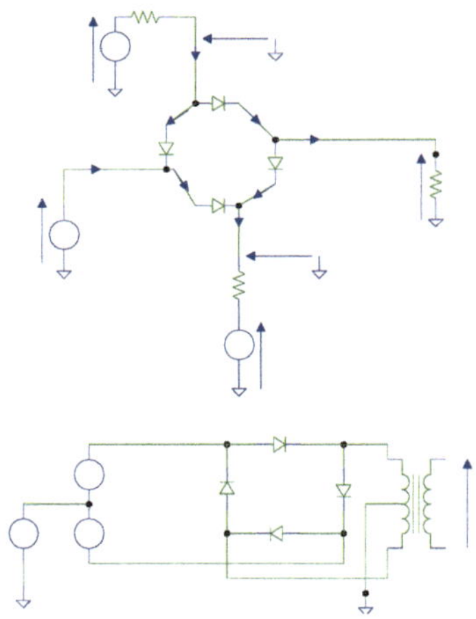

Figure 1.10. *Analog switch and ring modulator*

1.4. Detailed model of pn diode: variable conditions and effect of temperature

1.4.1. *Junction at equilibrium (at zero voltage or in open circuit): diffusion potential*

If the p zone is doped with N_A acceptors per volume unit and the n zone with N_D donors per volume unit, the carrier concentrations in neutral zones are: $p = N_A$ on the p side and $n = N_D$ on the n side in semiconductors with a small forbidden bandgap such as Si and GaAs.

A space charge zone (thickness W_A on side p, W_D on side n, total W_T) is created at the junction by recombination of a certain quantity of electron-hole pairs. The fixed charges that appear then balance: $N_A W_A = N_D W_D$, giving rise to an electric field $E(x)$ (algebraic in Figure 1.11) obtained by integrating the Poisson equation $\mathrm{div}(\varepsilon E(x)) = \rho(x)$, that is, to a single dimension x: $\frac{dE(x)}{dx} = \frac{\rho(x)}{\varepsilon}$, where ρ is the volumic charge density and ε the absolute dielectric permittivity.

If we disregard the mobile charge density in the space charge zone, we obtain a negative electric field (essentially directing the + fixed charges toward the – fixed charges) with linear variation according to the abscissa in each p and n part, respectively (Figure 1.11).

By integrating this electric field, we obtain the variation of internal potential Φ between regions p and n (not measurable on the exterior of the diode as it is compensated by the differences in contact potential V_{c1} and V_{c2}, which verifies $\Phi = V_{c1} + V_{c2}$ in such a way that no current circulates at equilibrium).

However, it is the absence of total current (both of holes and electrons) when the diode is in open circuit, without any potential difference applied by an external source, that fixes the total value of the internal potential difference Φ, called diffusion potential (and consequently, the width of the space charge zone $W_T = W_A + W_D$).

Indeed, here we need: $J_n = en\mu_n E(x) + \mu_n kT \frac{dn}{dx} = 0$ and $J_p = ep\mu_p E(x) - \mu_p kT \frac{dp}{dx} = 0$, where only components J_n and J_p are considered current densities along the Ox axis.

By accounting for $E(x) = -\frac{dV(x)}{dx}$, the equation above, for instance for holes, leads to:

$$-ep\frac{dV}{dx} = kT\frac{dp}{dx} \quad \text{or:} \quad -\frac{dV}{V_T} = \frac{dp}{p}$$

Hence: $\dfrac{p(x_2)}{p(x_1)} = \exp\left(\dfrac{V(x_1) - V(x_2)}{V_T}\right)$

provided that x_1 and x_2 are within the limits of the space charge zone submitted to the electric field $E(x)$.

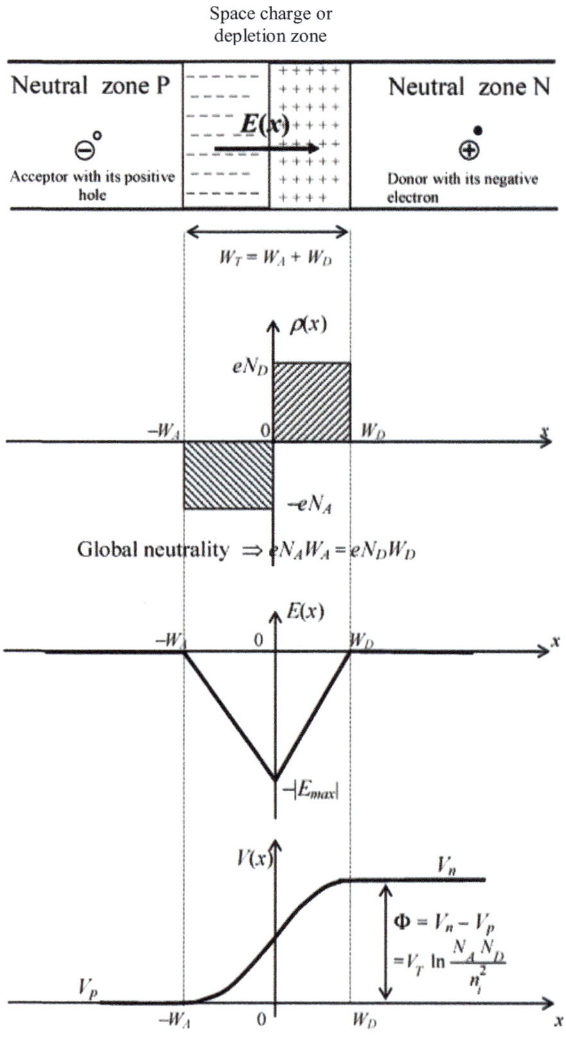

Figure 1.11. *Charges, algebraic electric field E(x) and potential in the pn diode at equilibrium (N.B.: the real field is in the inverse direction of the arrow, because negative E(x))*

At equilibrium (that is under zero external voltage):

$$\frac{N_A}{\left(\frac{n_i^2}{N_D}\right)} = \frac{N_A N_D}{n_i^2} = \exp\left(\frac{\Phi}{V_T}\right) \Rightarrow \Phi = V_T \ln\left(\frac{N_A N_D}{n_i^2}\right).$$

The built-in potential Φ represents a potential barrier for majority carriers and the electric field derived from this potential repels them into each neutral zone.

EXERCISE.– Calculate Φ for $N_A = 2 \times 10^{16}$ cm^{-3}, $N_D = 10^{18}$ cm^{-3}.

1.4.2. *Biased junction (or out of equilibrium): effect of an external potential difference*

Under an applied bias V_d, relation $\frac{p(x_2)}{p(x_1)} = \exp\left(\frac{V(x_1) - V(x_2)}{V_T}\right)$ still applies but we must account for the modification of internal potential by the potential difference V_d applied by the external source on the anode (P) relative to the cathode (N).

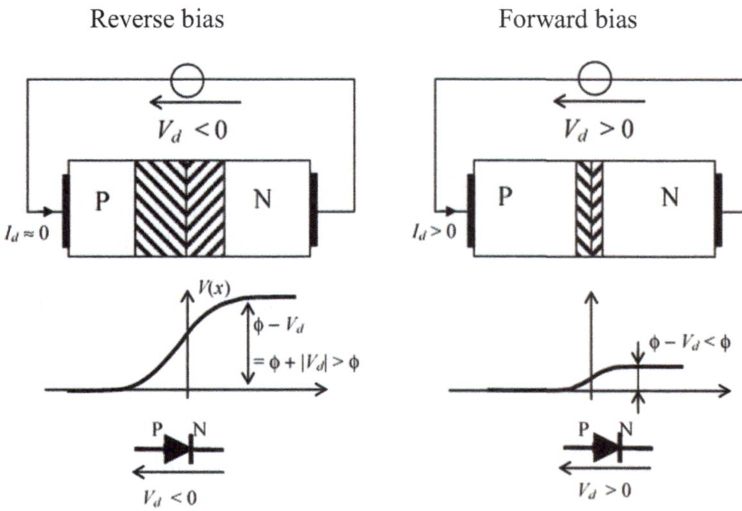

Figure 1.12. *Effect of external bias on charges and potentials in a diode (depletion zones hatched)*

We must then replace Φ with $\Phi - V_d$, where V_d represents the voltage applied by the external source on the p side relative to n side. Indeed, since the differences in contact potential V_{c1} and V_{c2} remain the same, the closed-loop equation becomes $\Phi - V_d = V_{c1} + V_{c2} - V_d$. The potential barrier is then reduced when V_d is positive (forward bias) and increased when V_d is negative (reverse bias). In the first instance, this has the effect of favoring diffusion currents for majority carriers through the space charge zone, with the electric field not being strong enough to repel them.

These flows of majority carriers correspond to both the migration of electrons through the type p neutral zone and to the migration of holes into the type n neutral zone. These carriers then become minority carriers at the entry to these neutral zones, in $-W_A$ and W_D, respectively, with $\exp(V_d/V_T)$ concentrations higher than at equilibrium.

The direct current is linked to this increase in minority carriers at the boundaries between space charge zone and neutral zones and may be calculated by accounting for the replacement of Φ by $\Phi - V_d$ and the value of Φ calculated earlier at equilibrium. For example, for holes whose concentration is fixed at N_A on abscissa $x_2 = -W_A$, we see the new concentration under bias V_d at the entry to the neutral zone of type n at $x_1 = W_D$:

$$\frac{p(W_D)}{p(-W_A)} = \exp\left(\frac{-(\Phi - V_d)}{V_T}\right) \quad \text{or} \quad p(W_D) = N_A \frac{n_i^2}{N_A N_D} \exp\left(\frac{V_d}{V_T}\right) = \frac{n_i^2}{N_D} \exp\left(\frac{V_d}{V_T}\right)$$

This concentration is higher than the minority carrier concentration at equilibrium n_i^2/N_D when V_d is positive (forward bias) and lower when V_d is negative (reverse bias). We will see an identical relation for electrons with the concentration n_i^2/N_A preceding factor $\exp(V_d/V_T)$. The diffusion currents are $\mu_n kT \dfrac{dn}{dx}$ and $-\mu_p kT \dfrac{dp}{dx}$, and become more significant in neutral zones in which the electric field is negligible in comparison to that in depletion zones due to a sufficient concentration of majority carriers to ensure conduction. $n(x)$ and $p(x)$ concentrations of minority carriers, which recombine in these zones with time characteristics τ_n and τ_p, are determined by recombination lengths L_n and L_p, linked to each τ and to the diffusion coefficients by $L_n = \sqrt{D_n \tau_n}$ and $L_p = \sqrt{D_p \tau_p}$. Since this will be considered in greater detail in Chapter 2, $n(x)$ and $p(x)$ verify the continuity equation, which accounts for the balance of mobile charges and is composed of a second-order differential equation. In a setting much greater in scale than L_n and L_p, $n(x)$ and $p(x)$ decrease exponentially according to the distance measured from the boundaries located at $-W_A$ and W_D with characteristic lengths L_n and L_p, known as recombination lengths. So, their derivative is simply $n(x)/L_n$ and $-p(x)/L_p$, with the signs indicating the decrease in $n(x)$ to negative x and $p(x)$ to positive x. Hence, a total diffusion current density that may be obtained by assessing $eD_n \dfrac{dn}{dx}$ at $x = -W_A$ and $-eD_p \dfrac{dp}{dx}$ at $x = W_D$, or:

$$J_d = e\left(\frac{D_n}{L_n}\frac{n_i^2}{N_A} + \frac{D_p}{L_p}\frac{n_i^2}{N_D}\right)\exp\left(\frac{V_d}{V_T}\right)$$

At equilibrium ($V_d = 0$), this current density is exactly compensated by the conduction current density $\underline{J_c} = e(n\mu_n + p\mu_p)\,\underline{E}$, which always has a negative sign in the circuit employed since the electric field is itself negative. So, for a diode of area A we have:

$$I_d(V_d) = Ae\left(\frac{D_n}{L_n}\frac{n_i^2}{N_A} + \frac{D_p}{L_p}\frac{n_i^2}{N_D}\right)\left[\exp\left(\frac{V_d}{V_T}\right) - 1\right]$$

with a preexponential factor $I_0(T) = Ae\left(\dfrac{D_n}{L_n}\dfrac{n_i^2}{N_A} + \dfrac{D_p}{L_p}\dfrac{n_i^2}{N_D}\right)$ proportional to n_i^2 and in which the currents of each type are in the same ratio to that of the two terms inside the brackets, that is $\dfrac{D_n}{D_p}\dfrac{L_p}{L_n}\dfrac{N_D}{N_A}$, also called the injection ratio, which can principally be adjusted by the ratio of doping concentrations on either side.

The reverse saturation current $-I_0(T)$ is obtained when the reverse bias is high enough to allow us to disregard the diffusion currents of the majority carriers. This conduction current of the minority carriers is proportional to the concentrations of the minority carriers, so n_i^2 / N_D and n_i^2 / N_A, for holes and electrons, respectively.

Since all of these currents are proportional to $n_i^2 = N_c N_v \exp(-E_g/kT) = BT^3 \exp(-E_g/kT)$ and to the area A of the diode; we define: $I_0(T) = A\left(\frac{T}{T_0}\right)^3 J_0 \exp(-E_g/kT)$, where J_0 is a current density independent of temperature.

Typically, this current is from several dozen pA to several nA in low-power silicon diodes.

Other phenomena become significant in real diodes, notably the recombination of electron-hole pairs in the space-charge zone, which was ignored previously. This results in the forward current varying less quickly with voltage than according to the ideal law, and $I_0(T)$ being higher than the value provided by the above-mentioned formula. We can correct for this by introducing the ideality factor η, typically taken between 1 and 2, to replace V_T with ηV_T. So, for the real diode:

$$I_d(V_d, T) = I_0(T)\left[\exp\left(\frac{V_d}{\eta V_T}\right) - 1\right]$$

The dynamic resistance of the diode will become: $R_d = \frac{\eta V_T}{I_d}$

EXERCISE.– Find the real current I_0 by measuring the variation factor of the current for a voltage variation of 0.1 V, the current value extrapolated to $V_d = 0$ and the ideality factor of diode 1N4148 (Figure 1.13) at $T = 300$ K.

In reverse bias, real diodes pass a current whose absolute value is higher than the theoretical value $I_0(T)$ calculated above and which increases with the reverse voltage. This effect is due to the extension of the depletion zone with the increase in the reverse voltage since the generation of electron-hole pairs, which is fostered by the presence of defects or recombinants centers in the depletion zone, occurs in greater volume. By integrating the generation rate in this volume, we can obtain a current that increases with reverse voltage, which we account for by an additional parallel conductance G in the equivalent circuit (Figure 1.14).

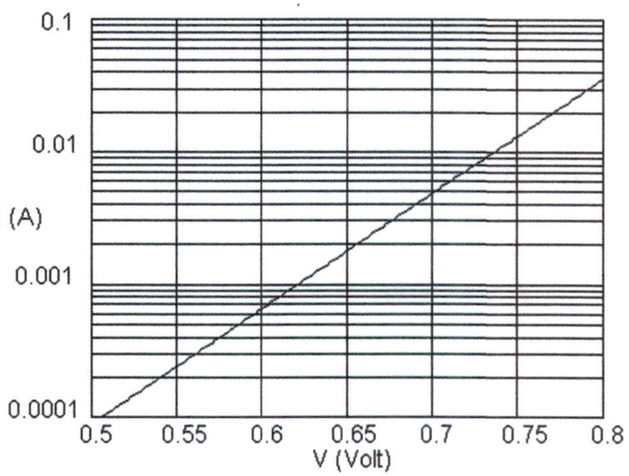

Figure 1.13. *Characteristic of diode 1N4148 in forward bias*

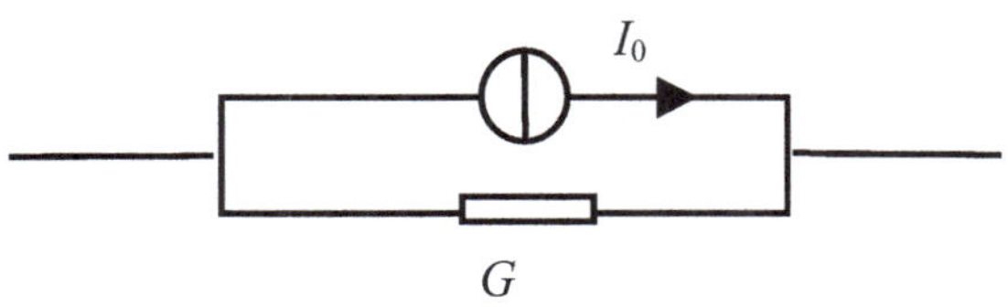

Figure 1.14. *Equivalent circuit of diode in reverse static bias*

Indeed, we should note that all real diodes under reverse voltage are limited to a value of V_{rmax} (absolute value of V_d). Once this voltage is reached, the electric field is strong enough to wrest electrons from their atoms (ionization), thereby creating further electron-hole pairs. In this manner, reverse current increases very quickly since the phenomenon becomes cumulative (avalanche breakdown). Dissipated power increases significantly, which can lead to the diode's immediate destruction if appropriate protection is not provided.

1.4.3. *Effects of temperature*

By deriving the logarithm of the preceding reverse current expression, we obtain: $\frac{\partial I_0(T)}{I_0(T)} = 3\frac{dT}{T} + \frac{E_g}{kT}\frac{dT}{T}$ corresponding to a 15% per degree increase at around 300 K, or equivalently a twofold increase approximately every 7° for silicon.

In forward bias, maintaining a current I_d constant and taking the logarithmic differentiation of I_d:

$$\frac{dI_d(T)}{I_d(T)} = 0 = \frac{\partial I_0(T)}{I_0(T)} - \frac{eV_d}{\eta kT}\frac{dT}{T} + \frac{e}{\eta kT}dV_d$$

Or, accounting for expression $\frac{\partial I_0(T)}{I_0(T)} = 3\frac{dT}{T} + \frac{E_g}{kT}\frac{dT}{T}$: $\left[3+\frac{E_g}{kT}\right]\frac{dT}{T} = \frac{eV_d}{\eta kT}\frac{dT}{T} - \frac{e}{\eta kT}dV_d$, which ultimately provides: $\frac{dV_d}{dT} = \frac{V_d}{T} - \frac{\eta k}{e}\left[3+\frac{E_g}{kT}\right]$,

with $V_d = 0.85$ V, $T = 300$ K, $\eta = 1.3$, $E_g = 1.12$ eV (silicon), one obtains: $\frac{dV_d}{dT} = -2.3$ mV/° (value widely provided in devices documentation but different for 1N4148).

NOTE.– When supplied with constant current in forward bias, we see the diode's forward voltage decrease as its temperature increases.

The increase in reverse current and decrease in direct voltage together with the increase in temperature represent drawbacks that limit the operating temperature of silicon components to around 200 °C. The only advantage is that the pn diode can be used as a temperature sensor.

Despite the increase in temperature coefficients $\frac{\partial I_0(T)}{I_0(T)}$ and $\frac{dV_d}{dT}$ with E_g, it is worthwhile to make use of semiconductors with larger forbidden bandgaps than silicon ($E_g = 1.12$ eV at 300 K) for components expected to operate at higher temperatures, such as SiC ($E_g = 2.8$ at 3.3 eV depending on crystal type), GaN (3.4 eV) or diamond (5.5 eV) since the tremendous decrease in $I_0(T)$ easily compensates for the increase in temperature coefficients.

1.4.4. *Capacitive effects*

If we apply the Gauss theorem to both sides of the space charge zone, then we can calculate the maximum electric field, whose absolute value E_{max} is situated on abscissa 0, and whose algebraic value is negative when accounting for the conventions in Figure 1.11:

$$E_{\max} = \frac{e N_A W_A}{\varepsilon} = \frac{e N_D W_D}{\varepsilon}$$

which indicates that there is a charge per unit area identical in absolute value $e\,N_A\,W_A = e\,N_D\,W_D$ on each side of the charge space zone. By accounting for $E(x) = -dV/dx$, by integration we obtain the difference of internal potential $\Phi - V_d$ between zones n and p, that is from the total area of the triangles representing $E(x)$, $\frac{1}{2}(W_A + W_D)E_{\max}$ or rather $\Phi - V_d = \frac{\varepsilon}{2e}\left(\frac{1}{N_A} + \frac{1}{N_D}\right)E_{\max}^2$.

Hence, we deduce $E_{\max} = \sqrt{\frac{2e}{\varepsilon\left(\frac{1}{N_A} + \frac{1}{N_D}\right)}}\sqrt{\Phi - V_d}$ and the total width $W_T = W_A + W_D = \sqrt{\frac{\varepsilon}{2e}\left(\frac{1}{N_A} + \frac{1}{N_D}\right)}\sqrt{\Phi - V_d}$ of the charge space zone or depletion zone. The charge Q per unit area on either side of the charge space zone may then be easily calculated in absolute value by:

$$Q = e\,N_A W_A = e\,N_D W_D = \sqrt{\frac{2e\varepsilon}{\frac{1}{N_A} + \frac{1}{N_D}}}\sqrt{\Phi - V_d}$$

Technological limitations mean that we either have $N_A >> N_D$, or $N_D >> N_A$ in most cases, which often means we can disregard one of the two terms in the denominator.

One variation of dV_d leads to variation dQ, hence the capacitance per unit area for a reverse biased diode ($V_d < 0$) and in low forward voltages (<0.3 V) reads:

$$C_i = \left|\frac{dQ}{dV_d}\right| = \sqrt{\frac{e\varepsilon}{\frac{2}{N_A}+\frac{2}{N_D}}}\frac{1}{\sqrt{\Phi - V_d}}$$

Accordingly, this capacitance decreases as reverse voltage $-V_d = V_r$ increases, corresponding to an increase in the width of the depleted zone $W_T = W_A + W_D$. This same expression applies for the parallel plate capacitor, which has a capacitance per unit area of ε / W_T.

The equivalent reverse circuit is thus reduced to capacitance C_i if we can disregard the reverse current and conductance:

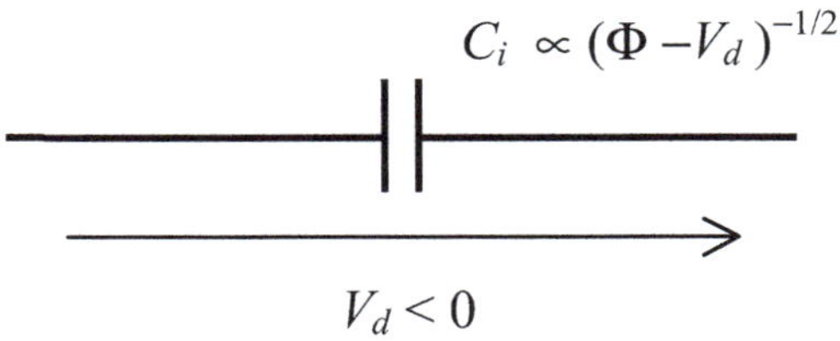

Figure 1.15. *Equivalent circuit for reverse biased diode*

EXERCISE.– Find the capacitance for a square diode of 1 mm width, then for one of 10 μm width if $N_A = 2 \times 10^{16}$ cm^{-3}, $N_D = 10^{18}$ cm^{-3} , $\varepsilon = \varepsilon_0\varepsilon_r$ ($\varepsilon_r = 11.7$ for Si), under $V_d = 0$ V then $V_d = -10$ V (use the value of Φ calculated earlier).

In forward bias, an additional charge Q_s of minority carriers is stored in the neutral zones, the injection and diffusion of these carriers being the very mechanism of forward current flow. In fact, this charge is neutralized by an equivalent charge of opposite sign due to the majority carriers that are attracted. While they do neutralize each other, these charges undergo variations when the applied voltage varies, which leads to a capacitive effect. Due to the exponential dependence of this charge Q_s and current I_d with forward voltage, there is proportionality between them, so: $Q_s = \tau I_d$ in steady state.

The time constant $\tau\,(=\tau_p$ or $\tau_n)$ is the average lifetime of minority carriers (from several ns to several µs) in the presence of majority carriers in concentration N_D or N_A. This is linked to the diffusion and length coefficients by the relation $L_p = \sqrt{D_p \tau_p}$ and $L_n = \sqrt{D_n \tau_n}$ due to the connection between minority carrier recombination and diffusion, which is a result of their concentration gradient due to the disappearance of a fraction of these mobile charges through the recombination phenomenon as we move away from the depletion zones.

This leads to the existence of a capacitance called diffusion capacitance, with small signals around an operating point: $C_d = \dfrac{dQ_s}{dV_d} = \tau \dfrac{dI_d}{dV_d} = \dfrac{\tau}{R_d}$.

Hence, the complete equivalent circuit under forward bias (Figure 1.16), which also accounts for access resistance, known as R_s series resistance, which provokes an additional voltage drop so that the real voltage between the terminals of the diode is actually $V_d' = V_d + R_s I_d$.

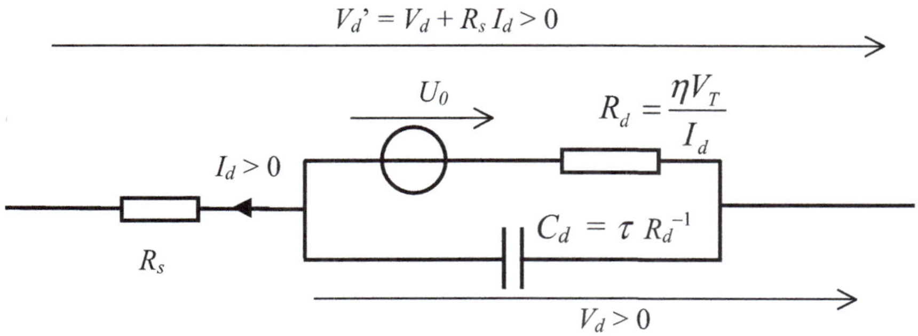

Figure 1.16. *Complete equivalent circuit of the pn diode in forward bias*

In any variable conditions: $I_d(t) = \dfrac{dQ_s}{dt} + \dfrac{Q_s(t)}{\tau}$.

This stored charge Q_s is the cause of the diode's frequency response and its delay in commutation (overlooking the connections' inductive effects here). The switching conditions are described below (Figures 1.17 and 1.18), assuming a much higher external series resistance R than the diode's internal resistance.

– On closure (flow open state → conductive state or "off → on"), the voltage at diode terminals does not immediately go from $-E_1$ to the forward voltage value U_0 since it must first discharge the reverse capacitance C_i and charge the diffusion capacitance C_d. At the terminals, the potential difference across the resistance R is here $E_2 - (-E_1) = E_2 + E_1$ at $t = t_0$ while it will become $E_2 - U_0$, which is lower, in permanent conditions; therefore, there is a current peak and a transient $v_D(t)$ of duration t_r to reach 0.9 U_0.

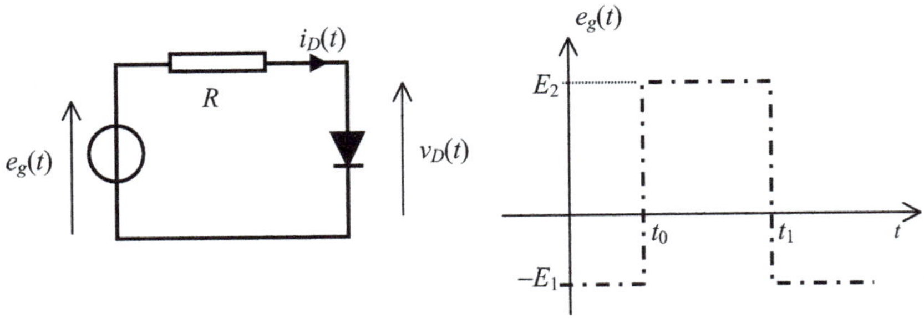

Figure 1.17. *Assembly for the diode test in switching conditions*

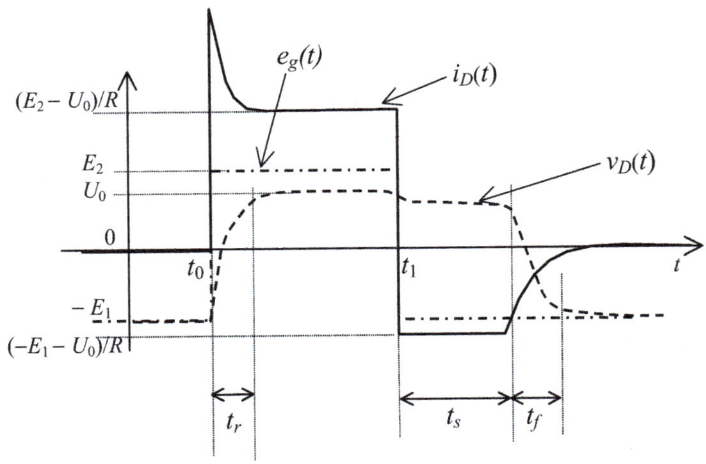

Figure 1.18. *Voltage and current in the diode in switching conditions*

– On opening (flow conductive state → open state or "on → off"), the diode imposes a voltage approaching U_0 due to the excess minority charges that arise during the storage time, which still corresponds to a conductive state. The current is then deduced from the drop in voltage across the R resistance, that is $(-E_1 - U_0)/R$, which is constant; it is a negative or reverse current that allows for the diffusion capacitance to be discharged. According to the Q_s equation in variable setting (first-order differential equation with constant second member; see exercise), Q_s decreases exponentially toward zero (duration t_s). Subsequently, capacitance C_i recharges under the reverse voltage $-E_1$ with time constant RC_i (duration t_f) by means of a reverse current tending finally to zero.

1.5. Different types of diode and their functions

1.5.1. *Zener diode*

The Zener diode is a diode in which the avalanche phenomenon is safely handled (non-destructive), allowing for a near constant reverse voltage $V_r = -V_d = V_Z$ to be maintained once attained (Figure 1.20). Dynamic resistance R_Z is in the order of several ohms to several hundred ohms and V_Z can be found between 1 and 200 V.

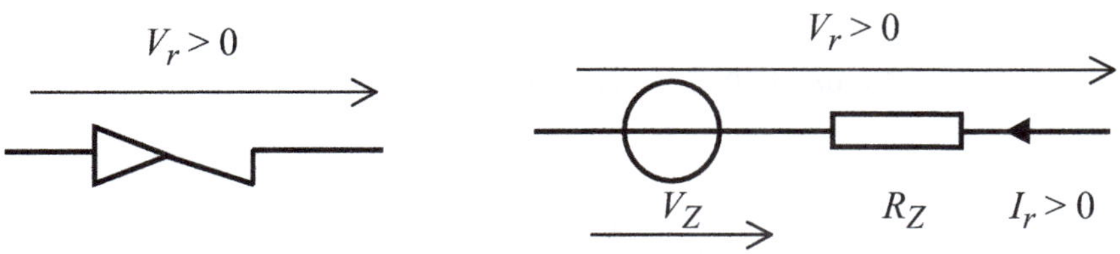

Figure 1.19. *Symbolic and electric representations of the Zener diode*

This diode allows us to make direct voltage stabilizers as well as variable voltage peak limiters.

1.5.2. *Schottky diode or metal–semiconductor diode*

In this diode, one side (p or n) of the semiconductor is replaced by a metal (Figure 1.21). The internal potential difference Φ is consequently lower, which results in a lower drop in forward voltage and a stronger reverse current than in a pn diode under the same conditions. Since less power is lost in forward bias, these diodes work to rectify strong currents (power diodes).

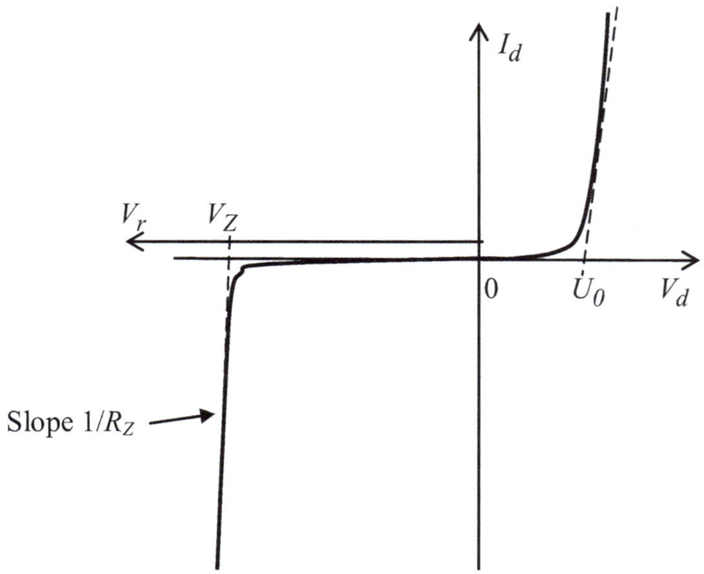

Figure 1.20. *Characteristic of Zener diode*

Figure 1.21. *Symbol of the Schottky diode or metal–semiconductor diode*

The metal injects a negligible quantity of minority carriers under forward bias, thus suppressing the stored charge Q_s. As a result, the switching times are significantly reduced and low-power diodes can have commutation times of less than a nanosecond, with operating frequencies up to 100 GHz possible.

1.5.3. *Light emitting diodes and laser diodes*

Electroluminescent diodes (light emitting diodes [LEDs]) are made exclusively with III–V semiconductors in which the phenomenon of recombination of electron-hole pairs is principally radiative: that is where the energy E_g recovered on the annihilation of two particles with opposite types is transformed into a photon of energy $h\nu$ approaching E_g and a wavelength close to $\lambda = \frac{c}{\nu} = \frac{hc}{E_g} = \frac{1.24}{E_g}$ (in µm if E_g is in eV).

Typically, these semiconductors are made up of the compound $GaAs_xP_{1-x}$, which allows for a forbidden bandgap of 1.8–2.6 eV, corresponding, respectively, to the colors red and blue (and at intermediate values corresponding to yellow and green) and more recently GaN and $In_xGa_{1-x}N$ for the near ultraviolet, purple and blue.

Laser diodes make use of the stimulated emission phenomenon that occurs in a resonant cavity, allowing for a highly directive luminous beam and very precise frequency (very tight frequency, unlike LEDs). By modulating the light beam, we can transmit information optically at high speed, typically using the infrared range (1.55 µm) in which attenuation in the optical fibers is minimal.

1.5.4. *Photodiodes and photovoltaic generators*

Light can generate electron-hole pairs by transferring the energy of a photon $h\nu$ to an electron of the valence band, making it enter the conduction band if $h\nu \geq E_g$ (by the opposite phenomenon to that at work in LEDs), thereby creating a hole in the valence band. If this phenomenon occurs in the space charge zone, the electric field separates the electron-hole pair, which gives rise to a current of the same type as the reverse current and which corresponds to a downwards shift in all $I_d(V_d)$ characteristics (Figure 1.22).

This current, called a photocurrent, is proportional to the flow of photons reaching the charge space zone, and can be collected in an external circuit to act as a means of measuring light flow (electric transcription of coded optical information) or of transforming the energy in sunlight to electric energy (photovoltaic or solar cells). In electrical terms, this diode acts as a generator.

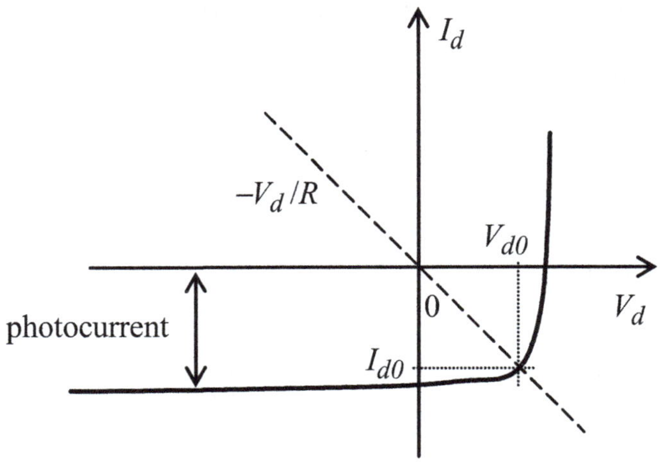

Figure 1.22. *Characteristic of a photovoltaic cell*

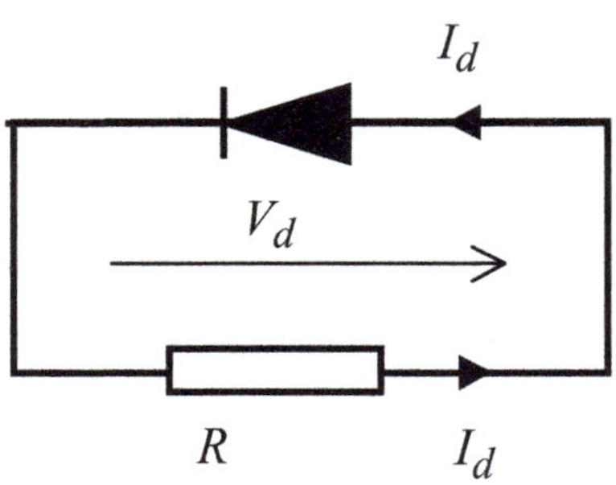

Figure 1.23. *Photovoltaic generator and its load resistance*

If we place resistance R between the terminals of the diode (Figure 1.23), the current flowing through it is $-\frac{V_d}{R}$, which is the equation of a straight line whose intersection with the photodiode's characteristic determines the operating point, with a current $I_{d0} < 0$ (Figure 1.22). In this quadrant, where $I_d < 0$ and $V_d > 0$, the diode is a generator due to it converting light energy into electrical energy. This is how photovoltaic cells operate. We can easily show that the power delivered in the load reaches maximum when the intersection of the straight line $-\frac{V_d}{R}$ with $I_d(V_d)$ occurs at the furthest point from the origin, situated in the elbow of its characteristic curve. Accordingly, the production of renewable electrical energy depends on this type of component, which can now provide yields approaching 20%. The peak power that can be obtained is determined by that of the sun, which reaches a maximum of around 1 kW/m^2.

1.6. Exercises

1.6.1. *Analyze the operation of the following circuits for a sinusoidal voltage generator and ideal diodes (a) with no threshold and (b) with a threshold U_0*

Perform a qualitative analysis of operations, assuming that for (b) the diodes' threshold U_0 is far below the maximum voltage supplied by the generator and that the frequency is of the same order as the reverse of the circuit's time constant, where there is one. Complete all the circuits by identifying the electrical quantities with name and arrow.

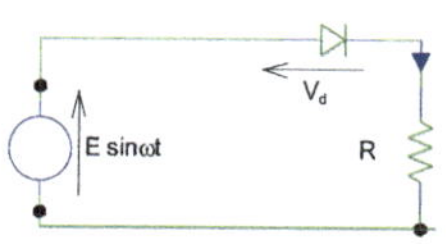

(i)

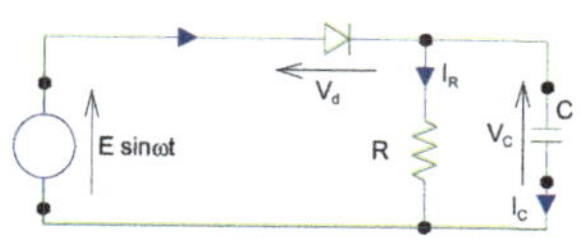

(ii)

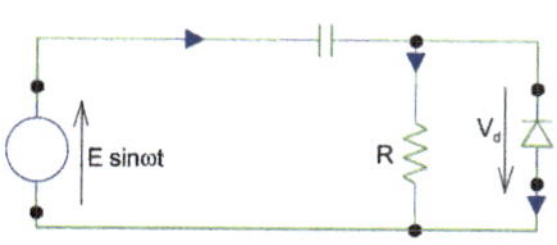

(iii)

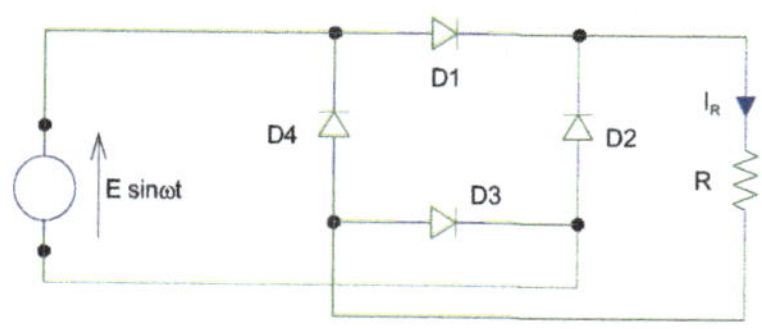

(iv)

Answer:

i) The diode is conductive when (a) Esinωt > 0 or (b) Esinωt – U_0> 0 and like an open circuit in other cases. When it is conductive, we find the voltage Esinωt or Esinωt – U_0 at the resistance terminals. When it has been opened, current is zero in resistance, as is voltage. This circuit acts as a half-wave rectifier.

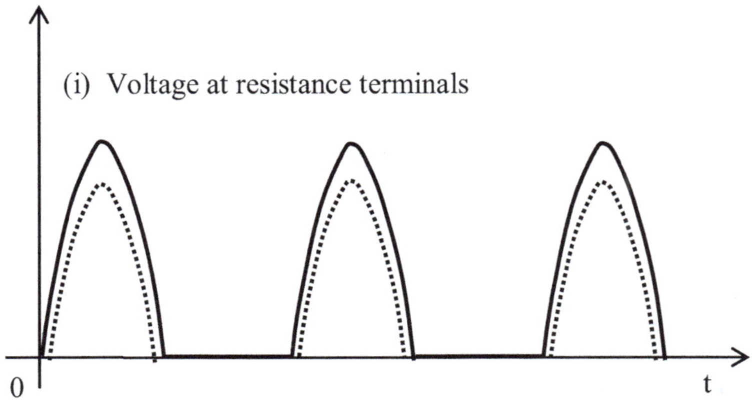

a) *Without diode threshold, solid line*

b) *With diode threshold, dotted line*

ii) As above, however this time the capacitor charges so that V_C becomes positive and remains so. The C capacitor discharges in the resistor when the diode is opened ($I_C < 0$), that is, if (a) Esinωt < V_C or (b) Esinωt – $U_0 < V_C$ and charges ($I_C > 0$) under voltage (a) V_C = Esinωt or (b) V_C = Esinωt – U_0 if the diode is conductive. This circuit acts as a filtered half-wave rectifier or works for signal envelope detection.

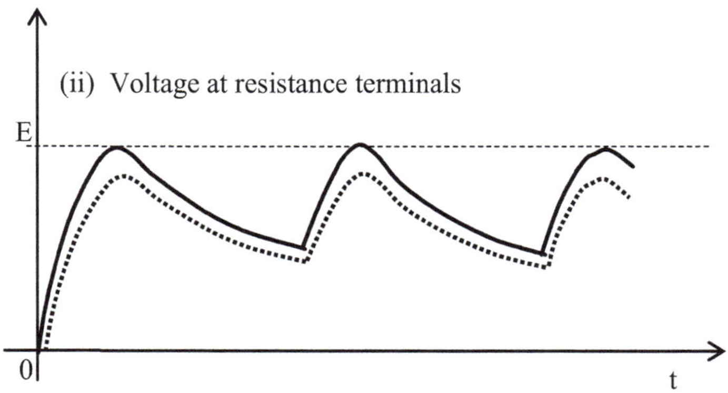

a) *Without diode threshold, solid line*

b) *With diode threshold, dotted line*

iii) Current I_d can only be negative inside the diode if voltage V_R at the resistance terminals is negative too, or zero if $V_R > 0$. During the two initial half-waves, current in the capacitor I_C is more negative than positive on average since for V_C we find (a) Esinωt or (b) Esinωt + U_0 when the diode is conductive, that is when Esinωt < 0 and (a) $V_R = 0$ or (b) $-U_0$, while we have only Esinωt – V_R when $V_R > 0$. Consequently, the capacitor charges to a mean value $V_{Cavg} < 0$. As V_R = Esinωt – V_C, this means $V_{Ravg} > 0$ since, the mean value of Esinωt being zero, we obtain, approximately, V_R = Esinωt – V_{Cavg} (with $V_{Cavg} < 0$). Hence, the voltage Esinωt is shifted upward except when the diode is conductive, that is when V_R becomes (a) = 0 or (b) = $-U_0$. The increase in voltage at the R terminals, caused by the initial charge of C, is required to attain stationary conditions in which the mean current becomes zero, since the capacitor would otherwise continue to charge indefinitely.

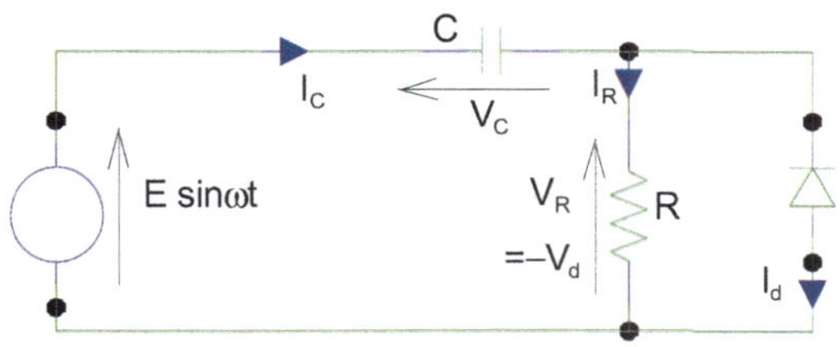

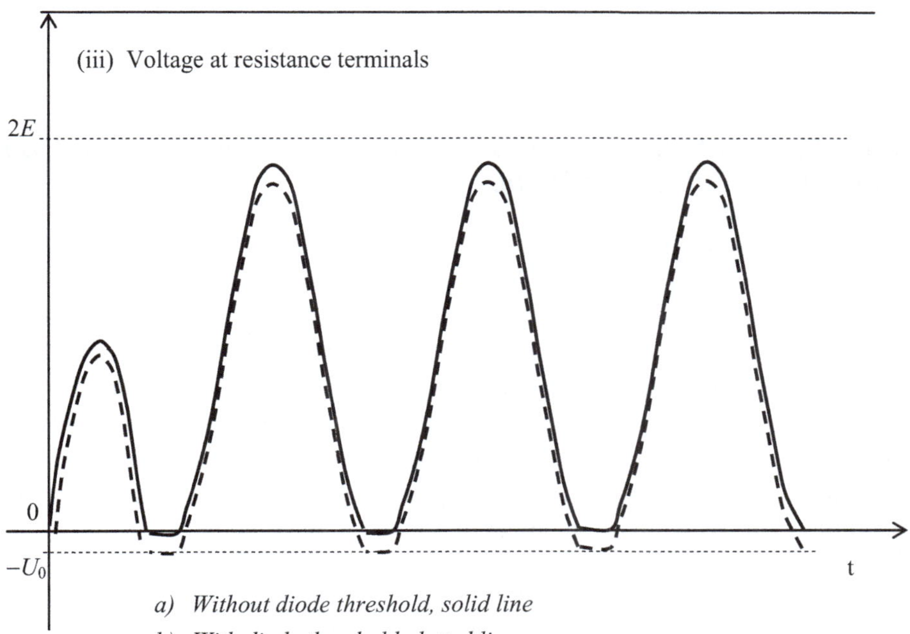

a) Without diode threshold, solid line

b) With diode threshold, dotted line

This circuit works to restore the signal's mean value by clamping, that is, forcing the most negative instantaneous value to become close to zero.

iv) This circuit operates as (i) but with D1, D3 conductive and D2, D4 opened when Esinωt > 0 and reverse when Esinωt < 0. The conductive diodes are always in series with the resistance here and the generator current is orientated toward resistance, so that I_R is always positive or zero. This is a full wave rectifier circuit.

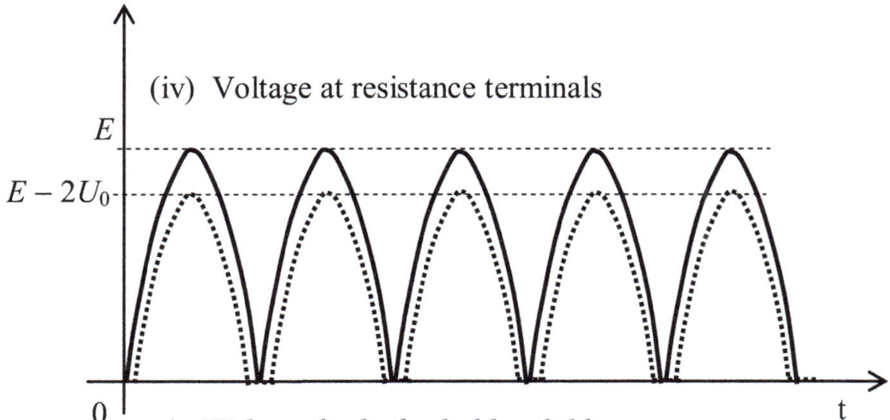

a) Without diode threhold, solid line

b) With diode threshold, dotted line

1.6.2. *Study of the current variations in a forward biased pn diode under influence of temperature*

Plot the forward characteristic $I_d(V_d)$ (at 300 K) of the 1N4148 diode with linear scales from 0 to 20 mA on a voltage scale from 0 to 1.2 V.

Find the literal expression of the load equation, that is, of $I_d = aV_d + b$ when a voltage generator E supplies the diode in forward bias, in series with an R resistance. Plot it for $E = 1$ volt and $R = 25\ \Omega$. Find the operating point (I_{d0}, V_{d0}), the threshold voltage U_0 and the dynamic resistance R_d of the diode from the plotted characteristic tangent at the operating point as well as the power dissipated in the diode.

By means of expressions in the course, use the provided, or calculated, diode characteristics and load resistance R to determine the relative variation of the current dI_d/I_d according to the relative variation dT/T of temperature. What happens if, by adjusting E so that current remains I_{d0}, R is equal to R_d? R becomes $<<R_d$? Conclude with the thermal stability of such a circuit.

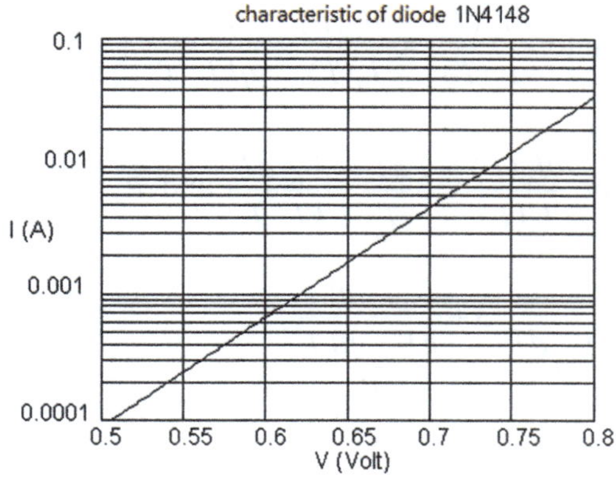

Answer:

Load equation: $E = V_d + R\,I_d$ or $I_d = \dfrac{E - V_d}{R}$

At the intersection, $V_{d0} = 0.74$ V and $I_{d0} = 10.3$ mA.

The plotted tangent gives a threshold of $U_0 = 0.68$ V and a dynamic resistance (the reciprocal slope of the tangent) of $R_d = 5.4\ \Omega$ (compare this with the theoretical value $\eta kT/eI_{d0} = 50$ mV/10.3 mA = 5 Ω). The dissipated power is $V_{d0}\,I_{d0} = 7.6$ mW.

By deriving the load equation,

we obtain: $\dfrac{dI_d}{I_d} = -\dfrac{dV_d}{RI_d}$

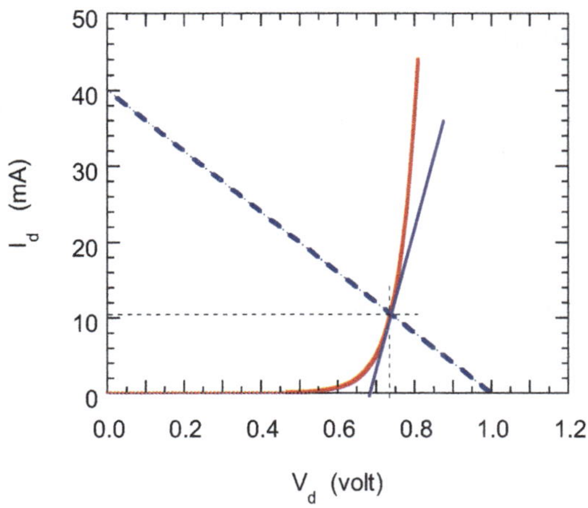

which we carry on to $\frac{dI_d(T)}{I_d(T)} = \frac{\partial I_0(T)}{I_0(T)} - \frac{eV_d}{\eta kT}\frac{dT}{T} + \frac{e}{\eta kT}dV_d$

and accounting for $\frac{\partial I_0(T)}{I_0(T)} = 3\frac{dT}{T} + \frac{E_g}{kT}\frac{dT}{T}$ and $R_d = \frac{\eta kT}{I_d}$

we obtain: $\frac{dI_d(T)}{I_d(T)} = \frac{dT}{T}\frac{3 + \frac{E_g}{kT} - \frac{eV_d}{\eta kT}}{1 + \frac{R}{R_d}}$

The numerator for silicon is 3 + 43 −15 ≈31 with E_g = 1.12 eV, kT = 26 meV and $\eta kT/e$ = 50 mV for the 1N4148 diode at ambient, deduced from the characteristic curve $I_d(V_d)$ in semilog, V_{d0} = 0.74 V. So for a variation of +10 °C from 300 K, $\frac{dI_d(T)}{I_d(T)} = \frac{10}{300}\frac{31}{1 + \frac{R}{R_d}} \approx \frac{1}{1 + \frac{R}{R_d}}$ which gives a 0.5 or 50% increase in I_d when $R = R_d$ and the relative variation approaches 100% if $R \ll R_d$. In this last instance, there is a danger of a cumulative effect, since the dissipated power increases, which contributes to further increase the component's temperature and thereby current, up to its destruction by overheating due to this thermal runaway.

Accordingly, it is crucial to maintain a load resistance $R > R_d$.

1.6.3. *Analog switch*

The diodes are identical, each having a conduction threshold of U_0 = 0.6 V, with negligible dynamic resistance and reverse current.

First case: V7 = −10 V; V8 = +10 V.

Find the state of the diodes when Ve = 0. Deduce Is and Vs. Between what boundaries of Ve does the diodes' state remain identical to the state above?

Second case: V7 = +10 V ; V8 = –10 V.

Find the state of the diodes when Ve = 0. Now find the relationship between Ve and Vs assuming that the diodes remain in this state when Ve ≠ 0. In what boundaries of Ve is this hypothesis verified? What happens if Ve leaves these boundaries, toward higher values for example?

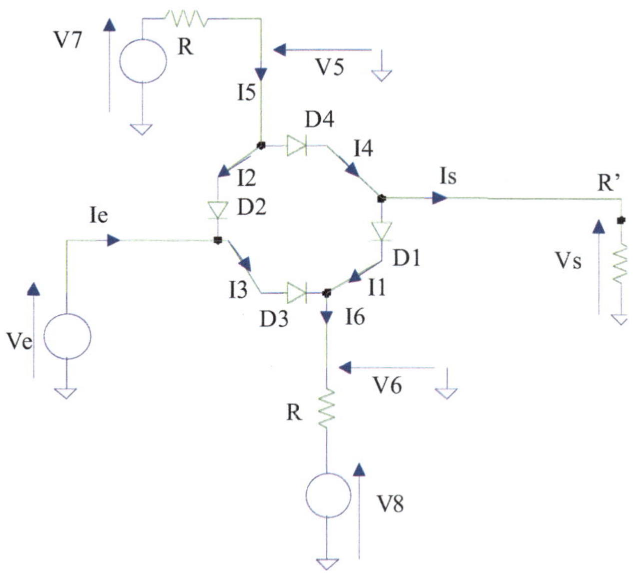

Answer:

First case: V7 = –10 V; V8 = +10 V; the loop equations going through V7, R, D2, Ve, on the one hand, and Ve, D3, R, V8, on the other hand, show that the diodes are open circuits when Ve = 0, so I5 and I6 are zero too, which leads to V5 = V7 = –10 V and V6 = V8 = +10 V. As a result, diodes D4 and D1 are also open circuits, which leads to Is = 0 and so Vs = 0. This remains true so long as the value of Ve does not change the state of the diodes: D2 only becomes conductive if Ve < –10.6 V and D3 only becomes conductive if Ve > +10.6 V; this interval –10.6 V < Ve < +10.6 V thus defines the range at which the switch is an open circuit, isolating the output from the input.

Second case: V7 = +10 V; V8 = –10 V.

With Ve = 0 and still going through the two same loops, we must conclude that diodes D2 and D3 are conductive, so V5 = U_0 and V6 = – U_0. The value of these two voltages V5 and V6 thus imposes that diodes D4 and D1 are also in a conductive state. Going through loop Ve, D2, D4, Vs or Ve, D3, D1, Vs, the diode thresholds compensate for each other since we have Ve – U_0 + U_0 –Vs = 0, or Vs = Ve = 0. However, this equation applies so long as the diodes remain in a conductive state, so Vs = Ve even if Ve ≠ 0. The switch is in the closed state, that is output and input voltages are the same, and current can flow from input to output or in the reverse direction.

If Ve increases in positive values, current Ie will increase and lead I2 to decrease to the point of cancelling it, closing diode D2. In order to assess this boundary, let us superimpose case Ve = 0, V7 = +10 V; V8 = –10 V with case Ve > 0, V7 = V8 = 0 V. When Ve = 0 = Vs, current I5 is symmetrically shared between I2 and I4 since the diodes are identical and operate under the same voltages, and Is = 0 since Vs = 0; so it is (10V – U_0)/R. In case Ve > 0, V7 = V8 = 0 V, current I2 should become a reverse current but we know that it will only subtract itself from (10V – U_0)/R when we superimpose two conditions; so we can continue to assume that diode D2 remains conductive with U_0 = V5 – Ve, and we will apply the same hypothesis to the state of other diodes. So, we must write the node equations, which provide:

$$\begin{bmatrix} 0 & 1 & -1 & 0 \\ 0 & -1 & 0 & -1 \\ -1 & 0 & 0 & 1 \\ 1 & 0 & 1 & 0 \end{bmatrix} \begin{bmatrix} I1 \\ I2 \\ I3 \\ I4 \end{bmatrix} = \begin{bmatrix} -Ie \\ -I5 \\ Is \\ I6 \end{bmatrix}$$

This system is of rank 3, as we see by replacing line 4 by the total of the 4 lines, and only allows a solution if we also have –Ie – I5 + Is + I6 = 0. In physical terms, this means that Ie has no choice other than to be equal to –I5 + Is + I6, which we know how to calculate according to the voltage drops in resistance. However, since there are only three independent equations for four unknowns, the system allows an infinite number of solutions for I1, I2, I3, I4, which can also not be known independently.

Nonetheless, we can consider the voltages, estimating that D2 will become an open circuit if Ve exceeds the value (V7 – U_0)R'/(R'+R) + U_0, which is obtain by making I5 = I4, I1 = 0 and Ie = I3. Indeed, two opposite diodes D2 and D1 will become open circuits for the same reason: too high a cathode potential relative to the anode; while on the other hand D3 and D4 remain conductive. In this case, the switch does not fulfill its role since Vs ≠ Ve.

1.6.4. *Ring modulator*

Use the same hypotheses as in the previous exercise. Let us assume that voltage V2 induced in the winding P2 when P1 is subjected to voltage V1 is equal to V1 and vice versa. The secondary S of the transformer provides voltage Vs = V1 + V2 so long as V1 + V2 are alternative symmetric.

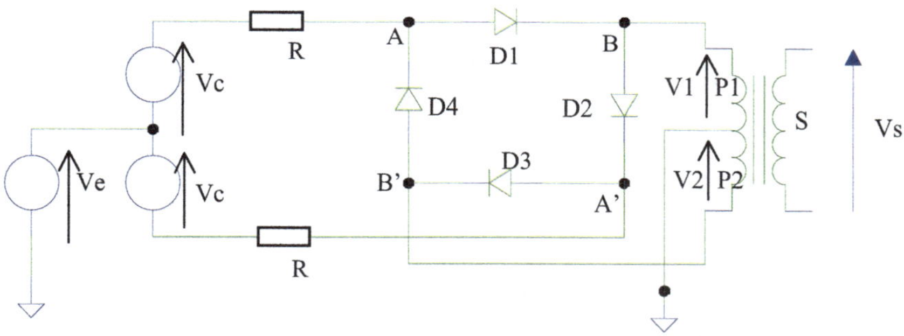

– Study in instantaneous values:

First case: Vc = +1.6 V.

Find the state of four diodes. Deduce from this the relation between Ve and V1 assuming that the current circulating in P1 does not disturb the state of the diodes. What condition must this current respect in order for this to occur? Where does it flow in a closed circuit?

Second case: Vc = –1.6 V. Same questions, replacing V1 by V2 and P1 by P2.

– Alternating Vc voltage: Vc is a periodic voltage of period T in symmetrical time levels of + or –1.6 V of equal duration to the half period. So, what is the shape of Vs and what are its limiting values? What happens if Ve is a voltage that varies much more slowly than T?

Answer:

– Study in instantaneous values:

First case: Vc = +1.6 V.

By circulating in the loop composed of sources Vc, resistors R, D1 and D2, we see that diodes D1 and D2 are conductive, since they are subject to a positive voltage between anode and cathode; and $2Vc - 2V_R - 2U_0 = 0$, so $V_R = Vc - U_0 = 1$ V. Consequently, D4 and D3 are open circuits and the potential of the low point of V2 (B') is therefore not imposed by these diodes.

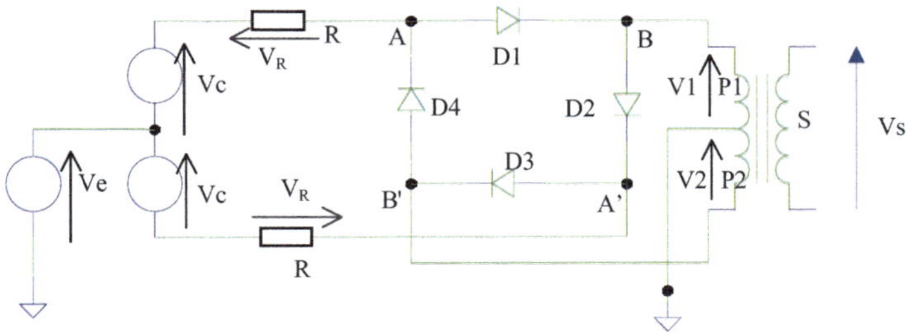

Accordingly, we again find that $V1 = Ve + Vc - V_R - U_0 = Ve$ at the terminals of P1 and V2 = V1 due to the coupling of two primary windings P1 and P2. The current output by Ve in P1 closes by ground, and is spread between arms Vc, R, D1 and Vc, R, D2 superimposing itself on the current already flowing into the loop composed of Vc, resistors R, D1 and D2. So, in absolute value, this must be less than $(Vc - U_0)/R$ in order not to induce blocking of one of the two diodes.

Second case: Vc = –1.6 V.

When the Vc sources are negative, D3 and D4 are conductive, hence $2Vc - 2V_R + 2U_0 = 0$ and D1 and D2 are opened. So, we have $V_R = -1$V and $-V2 = Ve + Vc - V_R + U_0 = Ve$; hence $V2 = -Ve = V1$, D1 and D2 indeed being open circuits.

– Alternating Vc voltage: According to the results above, Vs = V1 + V2 is alternatively 2Ve and –2Ve. So, Vs is a symmetrical alternative voltage of 2Ve maximum amplitude, which amounts to the product of Ve by a rectangular signal of ±2 value. The slow variable voltage Ve may be transformed into a much faster varying alternative voltage between ±2Ve, which is equivalent to modulating the amplitude of a symmetric rectangular signal.

1.6.5. *Switching diode: study on opening*

In forward bias and any given conditions, the (excess) minority stored charge is $Q_s(t)$ and is linked to direct current $I_d(t)$ by $I_d(t) = \frac{dQ_s}{dt} + \frac{Q_s(t)}{\tau}$, where τ is the lifetime of the minority carriers. The diode model in forward bias is reduced to the threshold voltage $U_0 = 0.75$ V.

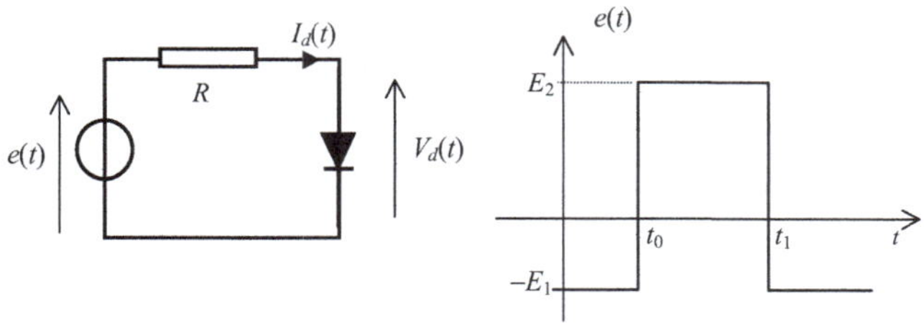

– Static conditions in forward bias (state "on" or closed switch): Calculate current $I_d(t) = I_{d0}$ according to R, U_0, E_2. Deduce the stored charge Q_{s0} in these conditions according to I_{d0} and τ.

– Storage time t_s on opening:

At $t_1 = 0$, we apply $e(t) = -E_1$. Assuming that the stored minority charge continues to impose $V_d(t) = U_0$, find current $I_d(t) = I_{d1}$. Next, solve the differential equation to obtain $Q_s(t)$ according to τ, E_1, E_2, R and t. Deduce from this the storage time t_s that separates instant t_1 from the instant when $Q_s(t)$ is cancelled out. How will this evolve with E_1 and E_2 (compare with the cases of $E_1 = 0$, $E_2 = 5U_0$ and $E_1 = E_2 = 5U_0$ for instance)?

– Fall time t_f of $V_d(t)$ on opening:

Following the cancellation of Q_s, the diffusion capacitance is negligible and the diode can be replaced by its only depletion zone capacitance C_i that we will consider as approximately constant. $V_d(t)$ is no longer held equal to U_0 and can move toward its final value. Draw the circuit and write the differential equation of $V_d(t)$. Solve this from the instant $t_2 = 0$ when these conditions apply with $V_d(0) = U_0$. Deduce the fall time t_f that separates the instant t_2 from that where $V_d(t_d) = -0.9\,E_1$ if $E_1 \neq 0$ or $V_d(t_d) = 0.1U_0$ when $E_1 = 0$. How will this occur with E_1 evolution?

– Calculate t_f, t_s, E_2, then τ for the diode 1N4153 in which $C_i = 2$ pF, with $R = 100\ \Omega$, $I_{d0} = 10$ mA and $E_1 = 6$ V if we measure $t_f + t_s = 2$ ns.

Answer:

1) In static conditions and a conductive state, $I_{d0} = \dfrac{E_2 - U_0}{R}$ and $Q_{s0} = \tau I_{d0} = \tau \dfrac{E_2 - U_0}{R}$.

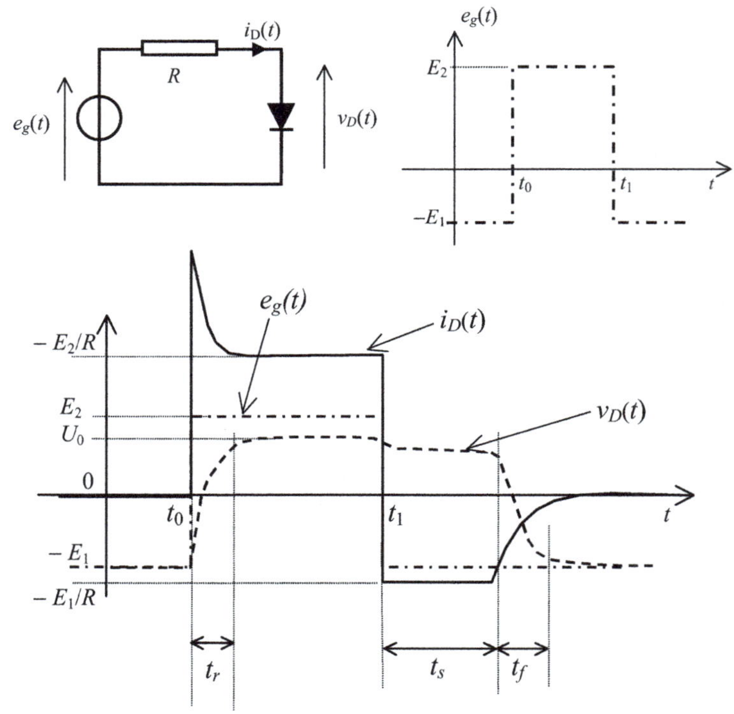

2) Switch conductive state → open state: at $t_1 = 0$, we apply $e(t_1) = -E_1$.

Assuming that the stored minority charge (which allows direct current to circulate), imposes $V_d = U_0$, providing $I_d = I_{d1} = \dfrac{-E_1 - U_0}{R} = -\dfrac{E_1 + U_0}{R}$; the differential equation is solved with $I_d(t) = I_{d1}$, and we write $Q_{s1} = \tau I_{d1}$: $\tau \dfrac{dQ_s}{dt} + Q_s = Q_{s1}$.

At $t = 0$, $Q_s = Q_{s0}$, or finally $Q_s(t) = (Q_{s0} - Q_{s1})\exp\left[-\frac{t}{\tau}\right] + Q_{s1}$;

$$Q_s(t) = \tau\frac{E_2 + E_1}{R}\exp\left[-\frac{t}{\tau}\right] - \tau\frac{E_1 + U_0}{R}.$$

From this, we may deduce the storage time t_s which goes from $t_1 = 0$ to the instant when $Q_s(t_s) = 0$, or

$$t_s = \tau\ln\frac{E_2 + E_1}{E_1 + U_0}$$

If $E_1 = 0$, $t_{s0} = \tau\ln\frac{E_2}{U_0}$; for example, for $E_2 = 5\ U_0 = 3.75$ V, $\ln\frac{E_2}{U_0} = 1.6$ while for $E_1 = E_2 = 5U_0$, $\ln\frac{E_2 + E_1}{E_1 + U_0} = 0.51$. So, we gain a factor 3 on t_s by using reverse voltage to switch the diode off, since there is a greater current allowing the stored charge to flow.

2) Ramp-down time of $V_d(t)$ between U_0 and $-0.9E_1$

Following the cancellation of $Q_s(t)$, the diffusion capacitance becomes negligible and we have the equivalent dynamic circuit below, where C_i is the capacitance of the space charge zone, assumed constant. The conditions below are represented with $t_2 = 0$, $V_d = U_0$.

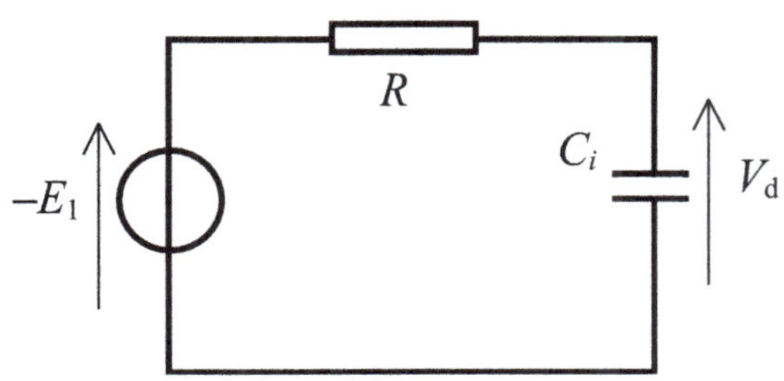

Hence, $C_i\frac{dV_d}{dt} = I_d(t) = -\frac{-E_1 - V_d(t)}{R}$, or rather $RC_i\frac{dV_d}{dt} + V_d(t) = -E_1$.

Accounting for the initial and final conditions, we obtain:

$$V_d(t) = (E_1 + U_0)\exp\left[-\frac{t}{RC_i}\right] - E_1 \text{ which provides for } V_d(t_f) = -0.9E_1.$$

Applying $V_d(t_f) = -0.9E_1$, $t_f = RC_i \ln\dfrac{E_1 + U_0}{0.1E_1}$

Numerical application for diode 1N4153: calculate $RC_i = 2 \times 10^{-10}$ s = 0.2 ns; $t_d = 0.2\ln\dfrac{6.75}{0.6} = 0.48$ ns; hence: 0.48 ns + t_s = 2 ns; then t_s = 1.52 ns and otherwise $E_2 = 100 \times 10^{-2} + 0.75 = 1.75$ V; $t_s = \tau \ln\dfrac{7.75}{6.75} = 0.138\ \tau$; hence τ = 11 ns.

NOTE.– According to the initial and final voltage values, the delay time can be either less or more than the respective time constants.

2

Bipolar Junction Transistors and Applications

2.1. The transistor effect

By the 1950s, the transistor effect had inspired the first semiconductor-controlled devices in which one electrode regulates the current flowing between two others. The effect relies on the injection of minority carriers into an electrode, called the "base", from the "emitter" electrode of the opposite type, in which the charge carriers are majority carriers. The third electrode, or the "collector", of the same type as the emitter, collects less current than that issued by the emitter but is dependent on the potential difference between base and emitter and on the emitter current. Let us consider the example of the pnp type, in which the base is p type. Historically, the term "base" was used due to the central position of this semiconductive part, which was originally used to diffuse doping impurities of the opposite type in order to create the emitter and collector on either side. Subsequent developments in semiconductor technology led to the production of bipolar junction transistors with architecture better matched to the regimes that favor the transistor effect, foremost of which were the control of base thickness and concentration of doping impurities in the collector. Indeed, base thickness affects the transistor effect by determining the percentage of emitter current that reaches the base–collector junction. The main factor affecting this current gain below 1 is due to the recombination of minority carriers in the base before they reach the depletion zone at the base–collector junction where they are aspirated by the electric field. We will consider another factor in this section covering the various types of current but, as a first approximation, if we model the bipolar pnp transistor on a single axis Ox as in the case of a pn diode, it can be represented in Figure 2.1.

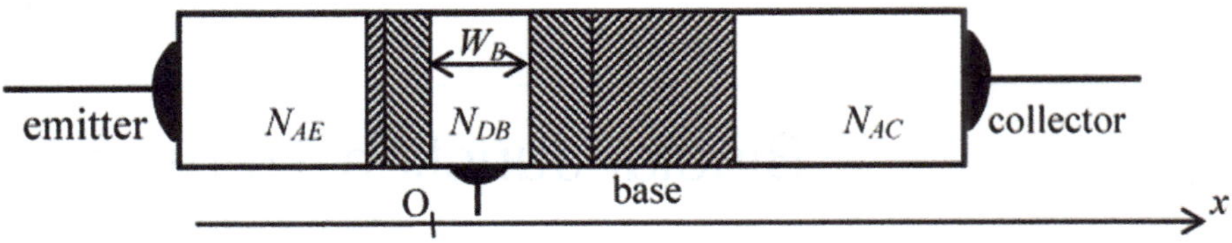

Figure 2.1. *One-dimensional geometry of a bipolar pnp transistor (depletion zones shaded)*

This has the appearance of two pn diodes upside-down and placed in series with the "base" electrode shared between them. The concentration of doping impurities is N_{AE}, N_{DB} and N_{AC} for the emitter (acceptors), base (donors) and collector (acceptors) respectively. Here, the base is a type n neutral zone of thickness W_B defined between the boundaries of the depletion zones at the emitter–base and base–collector junctions. The flow of holes that the emitter injects into the base is reduced by the phenomenon of recombination, since these holes are minority carriers in concentration $p(x)$ in the base and, in the presence of a far higher concentration of electrons approaching N_{DB} in the approximation known as low-level injection, valid so long as $p(x) << N_{DB}$, which we presume is verified here. This minority carrier flow leads to a diffusion current if we disregard the effect of the electric field, which remains weak in the neutral zones under low level injection. However, in addition to the current transport equations used in chapter 1, a continuity equation has to be verified for each type of carrier in concentration n or p. In order to write it, we will consider the flow balance of carriers entering and exiting a unit volume traversed by vector current densities J_n or J_p together with generation rates G_n or G_p and recombination rates R_n or R_p within this volume, that is, $\frac{dn}{dt} = G_n - R_n + \frac{1}{e} div(J_n)$ and $\frac{dp}{dt} = G_p - R_p + \frac{1}{e} div(J_p)$. In the base, we will only use the current of holes that gives rise to the transistor effect, which is written in one dimension and without a generation term (generation rates can be neglected in the dark and under low electric fields):

$$\frac{dp}{dt} = -R_p + \frac{1}{e}\frac{dJ_p}{dx}.$$

In the permanent regime $dp/dt = 0$ and in the low-level injection regime, the recombination rate is proportional to the difference in the minority hole population relative to that at equilibrium ($p_0 = n_i^2/N_{DB}$), with the reverse of the recombination time constant τ_p as proportionality factor: $R_p = \frac{p - p_0}{\tau_p}$.

The differential equation becomes $-\frac{p - p_0}{\tau_p} + \frac{1}{e}\frac{dJ_p}{dx} = 0$ or rather $-\frac{p - p_0}{\tau_p} + D_p \frac{d^2 p}{dx^2} = 0$ since the diffusion current is dominant and, as a consequence, we can disregard the conduction current. The solutions are composed of the sum of two exponentials with the coefficients depending on the two boundary conditions, themselves functions of potential differences V_{EB} and V_{CB}, such that $p(0) - p_0 = p_0\left[\exp\left(\frac{eV_{EB}}{kT}\right) - 1\right] = p_0 u_E$ and $p(W_B) - p_0 = p_0\left[\exp\left(\frac{eV_{CB}}{kT}\right) - 1\right] = p_0 u_C$ as in chapter 1. Writing the solution with hyperbolic functions gives:

$$p(x) - p_0 = \frac{p_0}{\sinh(W_B / L_p)}\left\{u_E \sinh\left(\frac{W_B - x}{L_p}\right) + u_C \sinh\left(\frac{x}{L_p}\right)\right\}$$

In order for the transistor effect to work, we must have $W_B << L_p$, which leads by developing at the first order to the linear relation $p(x) - p_0 = p_0 \frac{W_B - x}{W_B}$ with very acceptable accuracy. From the full expression, we can deduce the current density due to hole diffusion in the base:

$$J_p(x) = -eD_p \frac{d\left(p(x) - p_0\right)}{dx} = \frac{eD_p p_0}{L_p \sinh(W_B / L_p)}\left\{u_E \cosh\left(\frac{W_B - x}{L_p}\right) + u_C \cosh\left(\frac{x}{L_p}\right)\right\}$$

The loss of hole current by recombination in the base can then be calculated in relative value by:

$$1-\alpha_1=\frac{J_p(0)-J_p(W_B)}{J_p(0)}=\frac{(u_E+u_C)\left(\cosh(W_B/L_p)-1\right)}{u_E\cosh(W_B/L_p)-u_C}$$

where α_1 is the ratio of currents $J_p(W_B)/J_p(0)$ at base output and input.

As $W_B << L_p$, $\quad 1-\alpha_1\approx\frac{u_E+u_C}{u_E-u_C}\frac{W_B^2}{2L_p^2}$

Under a conductive regime, $u_E >> |u_C|$, which leads to $1-\alpha_1\approx\frac{W_B^2}{2L_p^2}$ or rather $\alpha_1\approx 1-\frac{W_B^2}{2L_p^2}$.

Injection ratio α_1 increasingly approaches 1 in inverse proportion to the base thickness, as it becomes smaller relative to the diffusion length of minority carriers in the base.

2.2. Bipolar junction transistor (or BJT) models and types

2.2.1. *Ebers-Moll model*

By accounting for all of the carrier flows in a transistor, together with the non-ideality of the EB and CB junctions, by their respective ideality factors η_E and η_C, we can describe the general operation as represented in Figure 2.2, while considering the forward gains α_F and reverse α_R that apply for minority carrier injection into the base by the emitter and collector, respectively. Hence, the total currents are given as:

$$I_C=-\alpha_F I_{EB0}\left[\exp\left(\frac{V_{EB}}{\eta_E V_T}\right)-1\right]+I_{CB0}\left[\exp\left(\frac{V_{CB}}{\eta_C V_T}\right)-1\right]$$

$$I_E=I_{EB0}\left[\exp\left(\frac{V_{EB}}{\eta_E V_T}\right)-1\right]-\alpha_R I_{CB0}\left[\exp\left(\frac{V_{CB}}{\eta_C V_T}\right)-1\right]$$

$$I_B=-(1-\alpha_F)I_{EB0}\left[\exp\left(\frac{V_{EB}}{\eta_E V_T}\right)-1\right]-(1-\alpha_R)I_{CB0}\left[\exp\left(\frac{V_{CB}}{\eta_C V_T}\right)-1\right]$$

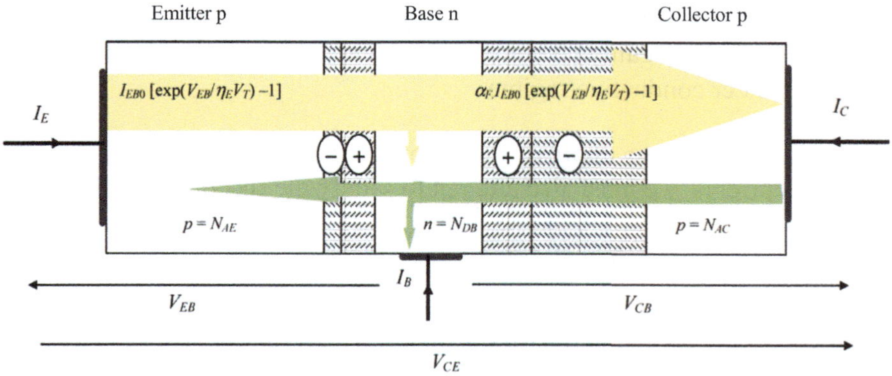

Figure 2.2. *Map of the various zones (depleted junction zones shaded) and currents in a pnp bipolar transistor*

In addition to the condition $W_B << L_p$ that maximizes the transfer of carriers from emitter to collector, the transistor effect depends on a second condition that helps minimize the majority carrier current (here, electrons) injected by the base in the emitter, that is $N_{AE} >> N_{DB}$.

Indeed, the ratio of hole current to total current in the EB junction is 1 minus the injection ratio of the two carrier types defined in chapter 1, where the diffusion length of the base is replaced by its thickness W_B that is much lower, that is $\alpha_2 = 1 - \frac{N_{DB}}{N_{AE}} \frac{W_B}{L_n} \frac{D_n}{D_p}$ with D_n and L_n as the diffusion coefficient and the diffusion length of the minority electrons in the emitter, respectively.

So, ratio α_F of the hole current reaching the collector to the total current in the emitter is $\alpha_F = \alpha_1 \alpha_2 = \left(1 - \frac{N_{DB}}{N_{AE}} \frac{W_B}{L_n} \frac{D_n}{D_p}\right)\left(1 - \frac{u_E + u_C}{u_E - u_C} \frac{W_B^2}{2L_p^2}\right) \approx 1 - \frac{W_B^2}{2L_p^2} - \frac{N_{DB}}{N_{AE}} \frac{W_B}{L_n} \frac{D_n}{D_p}$ since $u_E >> | u_C |$ under regimes in which the EB junction is under forward bias, which are those in which we attempt to evaluate the transistor effect.

This theoretical α_F ratio can be very close to 1, to as much as the nearest 10^{-2}, if $W_B << L_p$ and if $N_{DB} << N_{AE}$. It can be somewhat degraded by additional recombinations; these are, however, largely avoided in modern devices due to the improvement of semiconductors' crystalline quality. It is also called a current gain in a common base circuit, as specified below in the equations under conductive regime.

Finally, in order to improve transistor characteristics in off state or for very low currents, we must also attempt to satisfy the second inequality in:

$$N_{AE} >> N_{DB} >> N_{AC} \Rightarrow I_{EB0} >> I_{CB0}$$

This relation is also useful in making α_R much lower than 1 and thereby minimizing the reverse transistor effect, which can be considered as a parasitic effect. Keeping the pnp transistor as our example, we can introduce the following approximations into the previous expressions for currents under three operating regimes:

– *Conductive regime*: $V_{EB} > 0$ *and* $V_{CB} < 0$

$$I_C \approx -\alpha_F I_{EB0} \exp\left(\frac{V_{EB}}{\eta_E V_T}\right) - I_{CB0}$$

$$I_E \approx I_{EB0} \exp\left(\frac{V_{EB}}{\eta_E V_T}\right) + \alpha_R I_{CB0}$$

$$I_B \approx -(1-\alpha_F) I_{EB0} \exp\left(\frac{V_{EB}}{\eta_E V_T}\right) - (1-\alpha_R) I_{CB0}$$

As $I_{CB0} <<< I_{EB0} \exp\left(\frac{V_{EB}}{\eta_E V_T}\right)$: $I_C \approx -\alpha_F I_E - I_{CB0}$ and $\boxed{\frac{I_C}{I_B} \approx \frac{\alpha_F}{1-\alpha_F} = \beta}$

The current collector is proportional to the base current with a factor β, called current gain in a common emitter, somewhere between several dozen and several hundred. On the other hand, I_C is not subject to V_{CB} as a first approximation: $\Rightarrow$ *this is an ideal source of current.*

– *Off state:* $V_{EB} < 0$ *and* $V_{CB} < 0$

Only very weak reverse currents through the junctions remain: $I_C \approx \alpha_F I_{EB0} - I_{CB0}$; $I_E \approx -I_{EB0} + \alpha_R I_{CB0}$; $I_B \approx (1-\alpha_F) I_{EB0} + (1-\alpha_R) I_{CB0}$; that are leakage currents in an open switch.

– *Saturation regime*: $V_{EB} > 0$ *and* $V_{CB} > 0$

The two junctions are in forward bias: $V_{EB} = U_{0E}$ and $V_{CB} = U_{0C}$; $N_{AE} >> N_{DB} >> N_{AC} \Rightarrow U_{0E}$ (≈ 0.8 V) $> U_{0C}$ (≈ 0.7 V) $\Rightarrow V_{CE} = U_{0C} - U_{0E} < 0$; V_{CE} remains negative but very weak in absolute value.

$V_{CE} = U_{0C} - U_{0E}$ is equivalent to a near-zero voltage source, similar to a closed switch.

Under all regimes, the electric field at the CB junction retains the same direction and always draws carriers from the emitter to the collector.

NOTE.– For the npn transistor, all signs are inversed but the conclusions remain identical.

Accounting for the transistor effect, which offsets the characteristics of CB junctions due to current $\alpha_F I_E$ injected from the emitter, we can deduce the characteristics of transistors plotted according to the voltages referenced either at the base or at the emitter.

Pnp transistor:

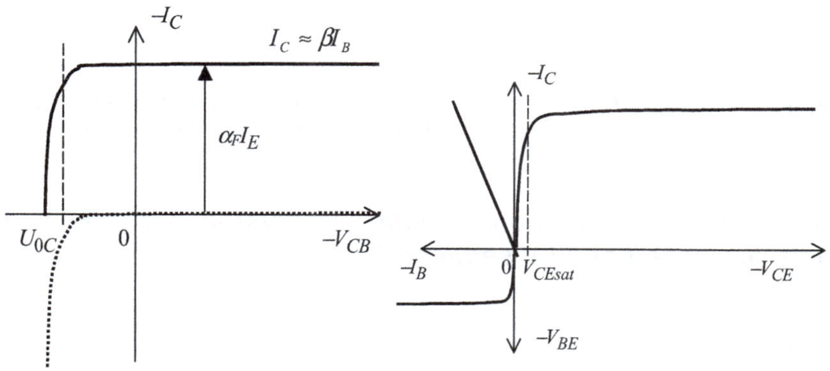

Figure 2.3. *Characteristics of pnp transistor with common base to the left (arrow α_F I_E represents the transistor effect) and with common emitter to the right*

Npn transistor:

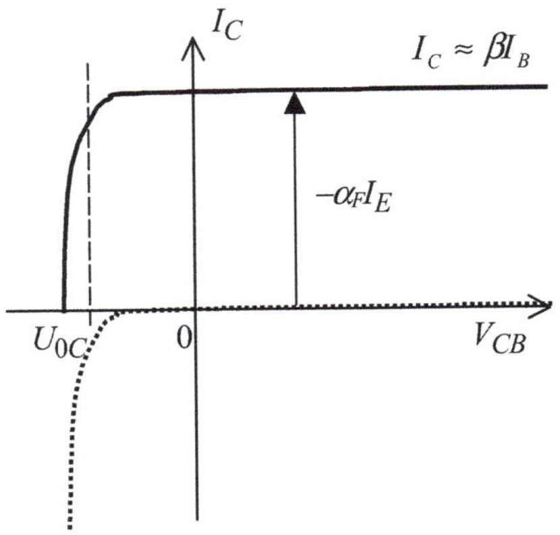

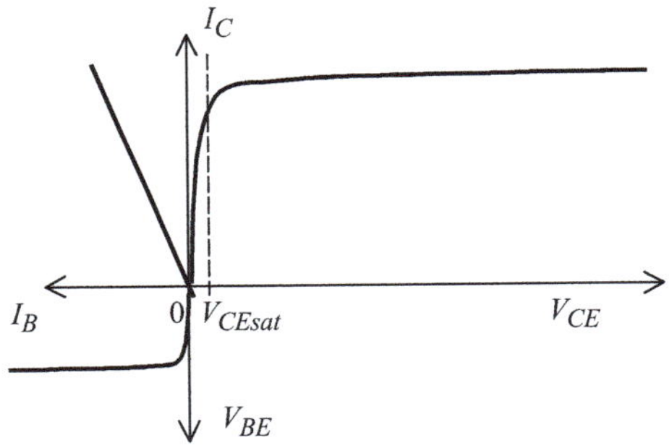

Figure 2.4. *Characteristics of npn transistor with common base to the left (arrow $-\alpha_F I_E$ represents the transistor effect) and with common emitter to the right*

2.2.2. The heterojunction bipolar transistor (or HBT) and transistors based on III-V semiconductors

Modern semiconductor material technology allows for the replacement of silicon in the base zone by an alloy of $Si_{1-x}Ge_x$ or $Si_{1-x-y}Ge_xC_y$, which has weaker forbidden bandgaps than silicon. This leads to the appearance of a potential barrier for the majority carriers in the base between base and emitter that significantly decreases α_R. We can then increase the doping concentration of the base, which has the effect of decreasing the associated time constants and significantly increasing operating frequency without loss of current gain β (the transition frequency for which $|\beta|$ is 1 then becomes several dozen gigahertz). Prior to the development of these alloys, fast bipolar transistors were made with III-V semiconductors in which (radiative) recombination is more than a thousand times faster than in silicon. For these materials, it was possible to provoke the appearance of the potential barrier for majority carriers in the base by adjusting the chemical composition of the alloys used in each zone. However, due to performance improvements and the ensuing increase in speed of field effect transistors, as studied in the next chapter, these older technologies have lost their speed and fallen into disuse, particularly for logic circuits of transistor–transistor logic (TTL) and emitter-coupled logic types.

2.3. Bipolar junction transistor in static regimes and applications of the exponential characteristic $I_C(V_{BE})$

2.3.1. *Equivalent circuits for the three operating regimes*

The equivalent circuits may be deduced from the previous equations and the reduction effect of base thickness W_B when the reverse bias of the *CB* junction increases, which has the consequence of increasing the absolute value of the current collector (Early effect) as α_F is brought closer and closer to one, its ideal value. This effect is accounted for by a parallel conductance at the source of the collector current.

For circuitry with a shared emitter (grounded emitter), we can represent the three operating regimes as follows.

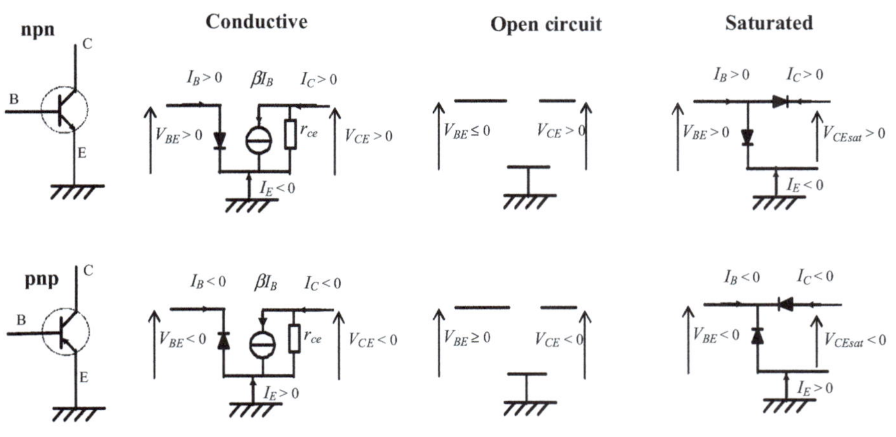

Figure 2.5. *Equivalent circuits of npn and pnp transistors under the three regimes (the diodes displayed by their symbol are under forward bias in terms of voltage, but the collector-base diode is still traversed by a reverse current due to the transistor effect).*

In the circuits mentioned above, diodes *CB* and *BE* are represented by  when under forward bias (despite being traversed by a reverse current in the case of the CB junction due to the transistor effect). Accordingly, we can either use the exponential expression of current for nonlinear applications or the equivalent linearized circuit (voltage source U_{0E} or U_{0C} and dynamic resistance r_{BE} = $\frac{\partial V_{BE}}{\partial I_B} = \frac{\eta_E V_T}{I_B} \approx \frac{V_T}{I_B}$) for linear applications.

2.3.2. Nonlinear applications: differential pairs and multiplier

Writing $I_{B0} = \pm(1-\alpha_F)I_{EB0}$ (+ for npn and – for pnp) and, for the sake of simplicity, taking $\eta_E = 1$, we have for the conductive regime ($V_{BE} > 0$ for npn and $V_{BE} < 0$ for pnp):

$$I_C \approx \beta I_{B0} \exp\left(\frac{\pm V_{BE}}{V_T}\right) \text{as soon as } |V_{BE}| > 3V_T$$

or rather $I_C = I_{C0}.\exp\left(\frac{\pm V_{BE}}{V_T}\right)$.

This characteristic is exponential.

Other functions can be achieved by associating two transistors as a differential pair.

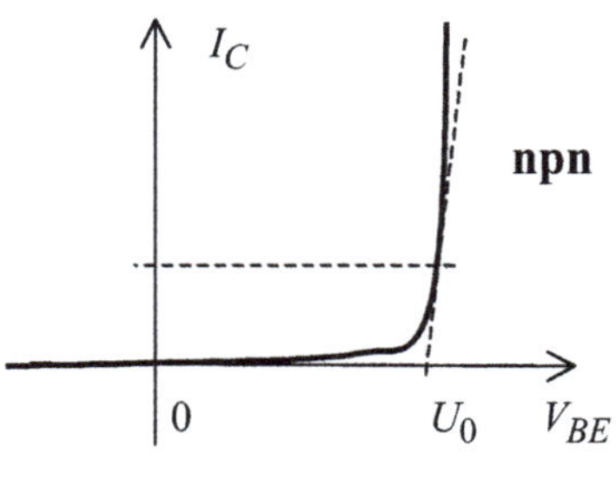

Figure 2.6. *$I_B(V_{BE})$ characteristic of npn transistor*

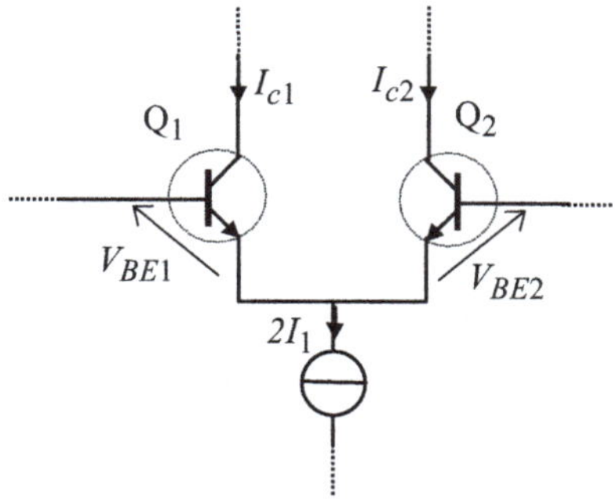

Figure 2.7. *Differential pair*

By decomposing each V_{BE} voltage according to differential voltage $V_D = V_{BE1} - V_{BE2}$ and common mode voltage $V_{MC} = \frac{V_{BE1} + V_{BE2}}{2}$:

$$V_{BE1} = \frac{V_{BE1} + V_{BE2}}{2} + \frac{V_{BE1} - V_{BE2}}{2} = V_{MC} + \frac{V_D}{2}$$

$$V_{BE2} = \frac{V_{BE1} + V_{BE2}}{2} - \frac{V_{BE1} - V_{BE2}}{2} = V_{MC} - \frac{V_D}{2}$$

By expressing currents I_{c1} and I_{c2} according to V_D and V_{MC}, we obtain:

$$I_{c1} = I_1 + I_1.\tanh\left(\frac{V_D}{2V_T}\right) \qquad \text{and} \qquad I_{c2} = I_1 - I_1.\tanh\left(\frac{V_D}{2V_T}\right)$$

Current variation is only linear for a range of several V_T around $V_D = 0$. In order to obtain a linear relation between currents I_{c1}, I_{c2} and V_D, we must modify the assembly as follows.

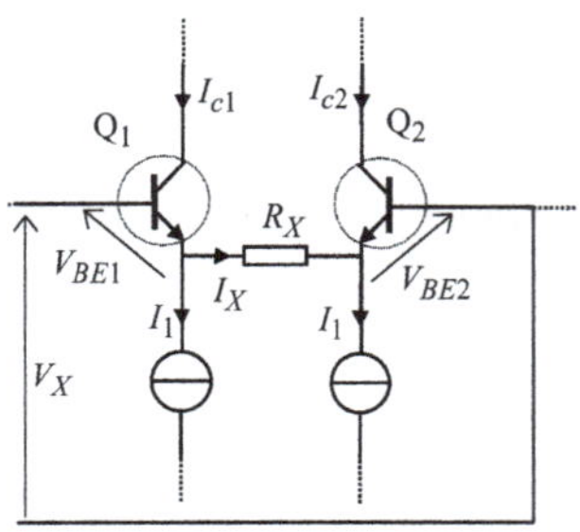

Figure 2.8. *Linearized differential pair*

Resistance R_X is taken as significantly higher than the dynamic resistance r_{BE} so that $V_X \approx R_X I_X$; hence $I_{c1} = I_1 + \frac{V_X}{R_X}$ and $I_{c2} = I_1 - \frac{V_X}{R_X}$.

The association of these two types of differential pair allows for the creation of an analog multiplier in which $I_A - I_B$ is proportional to the product of $V_X V_Y$ (see Figure 2.9 and exercise).

In this circuit, using both of the assemblies above, only "T" junctions represent a connection. The set-up operates a logarithmic transformation of the current difference proportional to V_Y with the Q_9-Q_{10} pair, providing the voltage V_D, and according to the function achieved by the double pair Q_5-Q_6-Q_7-Q_8, an anti-logarithmic transformation of the sum of V_D with the logarithm of the current difference proportional to V_X.

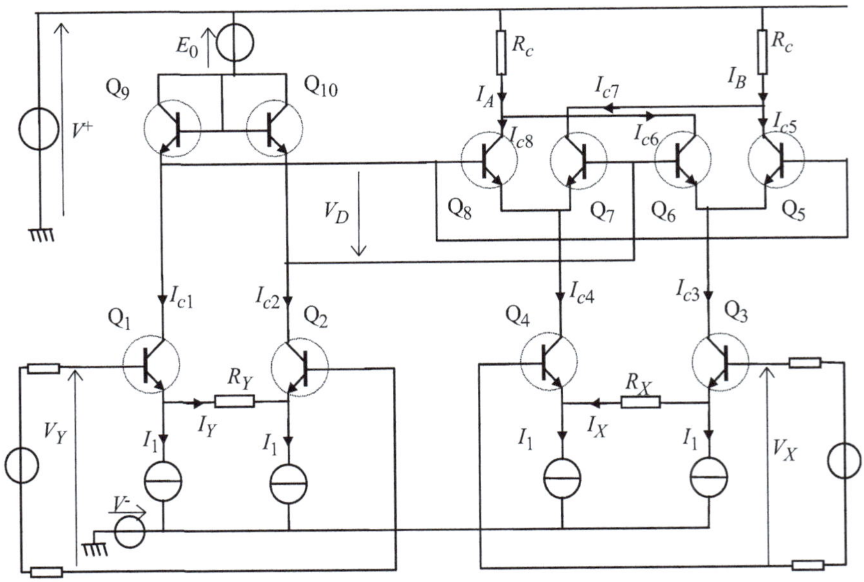

Figure 2.9. *Complete circuit of four quadrants multiplier*

2.3.3. *Circuits for transistor bias: current sources and current mirrors*

To define an operating point, we must first establish the transistor's voltage or static currents (or biasing currents). This is achieved by the following circuits:

– Resistance (Figure 2.10):

R_E and R_{B1} are optional but let us improve the circuit's thermal stability (see exercise on the effect temperature has on current in the pn diode, here BE, Chapter 1).

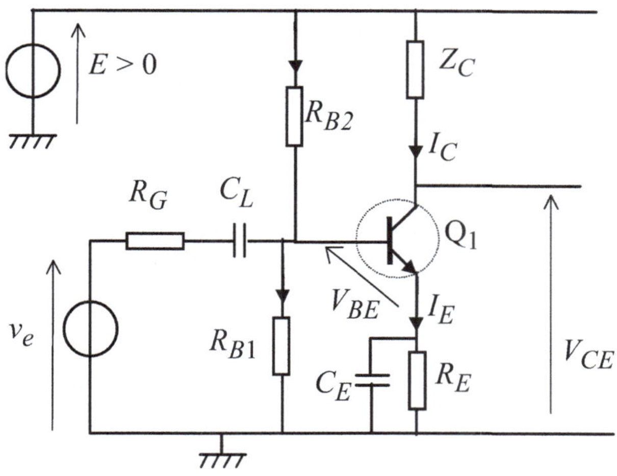

Figure 2.10. *Transistor biased by resistances*

The load impedance Z_C must comprise a parallel resistance in order to bias the collector. The capacitors have impedance that is negligible at the operating frequencies. This circuit is only used under high frequency and microwaves.

– Current source (Figure 2.11):

The current sources are indispensable in biasing the differential pairs and other circuits. In the circuit below, the Zener diode defines the sum of the base-emitter voltage and the voltage drop in the foot resistance R_E.

If U_{0E} is the threshold of diode BE, current I_C is given by: $I_C = \frac{V_Z - U_{0E}}{R_E}$.

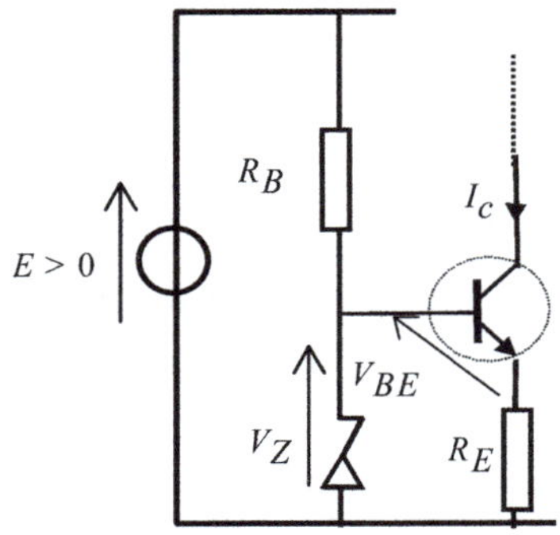

Figure 2.11. *Biasing by Zener diode and emitter resistance*

– Current mirrors (Figure 2.12):

The control current is I_C. If the transistors have identical characteristics, the equal BE voltages $V_{BE1} = V_{BE2}$ result in the equality of the collector currents $I_{C1} = I_{C2}$ and base current I_B.

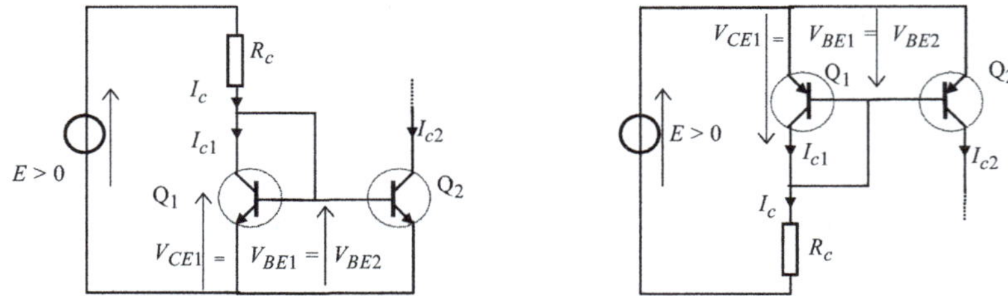

Figure 2.12. *Mirror circuit assemblies*

Here, $I_{C2} = I_C - 2\, I_B$, hence $I_{C2} = \frac{\beta}{\beta + 2} I_C$ very close to I_C.

We can modify the ratio I_{C2} / I_C by adding different resistances in series in the emitters or by placing n transistors, connected as Q_2, with each then sharing current I_{C2}/n rather than I_{C2} if they are identical. In integrated circuits, slave transistors such as Q_2 can have different sections to allow the proportionality coefficient I_C to be other than 1.

Here, $I_C = \frac{E - U_{0E}}{R_C}$, although this current can also be determined by the biasing current of another transistor: the collector current of another bipolar transistor or, more often, the drain current of a field effect transistor.

Biasing by the simple or multiple current mirror method is widespread in integrated circuits since it does not require resistance and uses only transistors (including a field effect transistor), which simplifies the technology.

2.3.4. *Voltage limitations*

Both junctions are subject to a voltage limitation, typically this is of a different nature depending on whether it is BE or CB.

In order to make the BE junction non-conductive, it is worthwhile to put it in reverse bias and thereby accelerate switching between the saturation or conductive regime and the blocking regime as we will see below, despite the fact that zero bias or a voltage just below the threshold is typically sufficient. However, this reverse voltage is limited to several volts and the limit seldom exceeds a dozen volts without running the risk of pushing the junction into avalanche regime. Indeed, as stated above, the injection of carriers from the emitter under conduction or saturation regimes requires the highest concentration of doping impurities in the emitter, then decreasing into the base followed by the collector. For this reason, the electric field, which triggers ionization by collision, and the ensuing phenomenon of carrier multiplication occurring in the emitter's depletion zone, is reached for lower reverse voltages than in all other zones of the transistor.

In contrast, due to the lower W_B thickness and limited doping of the base, the phenomenon of depletion zone extension for the CB junction, that may reduce W_B down to zero, is that which happens first when we increase the voltage applied to the collector. This effect, called the base punch-through, results in the sweep of carriers injected through the emitter by the electric field at the CB junction straight from the EB junction. Since the neutral part of the base is reduced to zero, there is no longer any current limitation either from the base control current or by recombination-diffusion and, as a result, the collector current increases abruptly without any noticeable rise in the collector voltage. Accordingly, the collector goes from being a current source to a voltage source as $|V_{CB}|$ (and consequently $|V_{CE}|$) does not exceed a certain limit, often referred to as V_{BR} (BR = breakthrough), as shown in Figure 2.13.

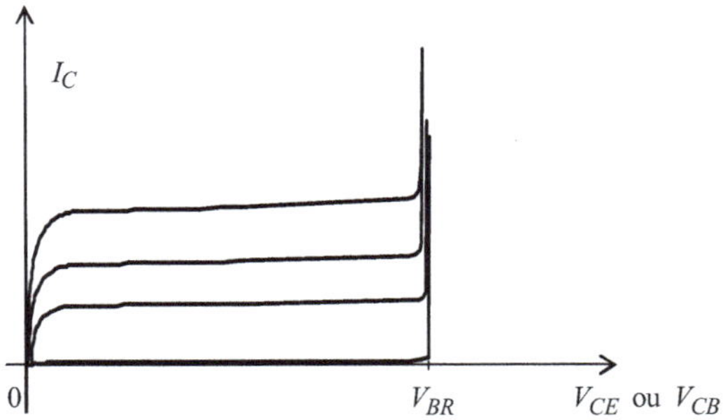

Figure 2.13. *Grid of characteristics $I_C(V_{CE}$ or $V_{CB})$ for an npn transistor at several base currents*

Current limitation must be anticipated in order to avoid transistor destruction due to overheating.

2.4. Small-signal dynamic circuits (linear approximation)

2.4.1. *Basic circuits*

If we make an approximation of a linear operation around an operating point, the BE diode voltage is represented by $V_{BE} = U_{0E} + r_{BE}\, I_B$, and the output collector current by $I_C = \beta I_B + \dfrac{V_{CE}}{r_{CE}}$. Indeed, the collector current increases with voltage V_{CE} for two reasons: first, the increase in V_{CB} leads to a decrease in base thickness W_B (see Figure 2.1), called the Early effect, since this is a reverse bias of the CB junction, which results in more effective injection due to an increase in α_F; second, in the case of a common emitter circuit, $V_{CE} = V_{CB} + V_{BE}$ also increases due to V_{BE} when we need to increase the collector current. Moreover, the load equation $V_{CE} = E - R_C\, I_C$ stands for the common emitter external circuit.

In dynamic equivalent circuits, the sources of biasing voltage E and U_{0E} are replaced by short circuits since their variations are zero. If we account for the state of diodes BE and CB, we can still consider $r_{CE} >> r_{BE}$.

Since the equations are linear, we can address the static values relative to the operating point (capital letters with index 1, 2, etc.) and small variations (lowercase) separately.

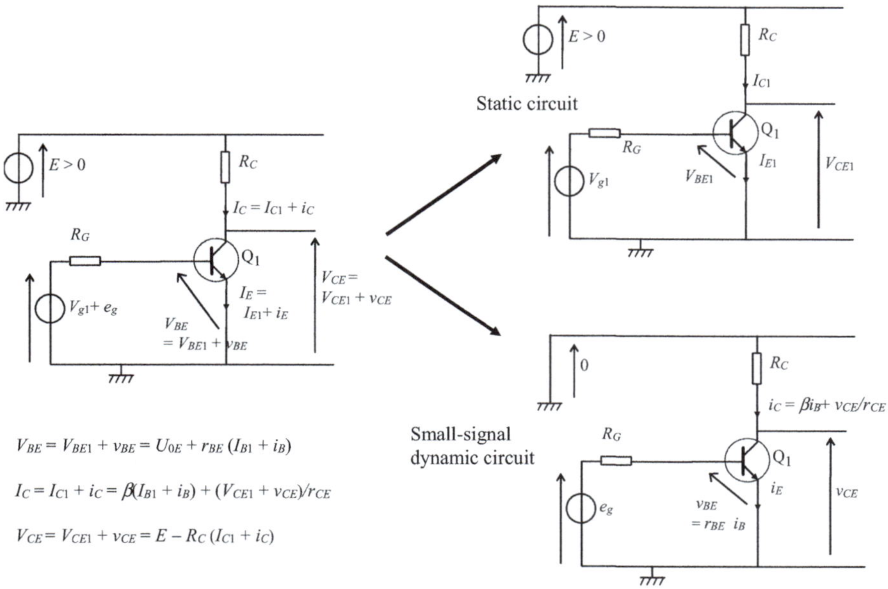

Figure 2.14. *Transition from schematic to static and dynamic electrical circuits*

The linearized small-signal circuit can be represented in Figure 2.15 (only the transistor is in the dotted rectangle).

Common emitter:

$$\begin{cases} v_{BE} = r_{BE} i_B \\ i_C = \beta i_B + \dfrac{v_{CE}}{r_{CE}} \end{cases}$$

or

$$\begin{cases} v_{BE} = r_{BE} i_B \\ i_C = g_m v_{BE} + \dfrac{v_{CE}}{r_{CE}} \end{cases}$$

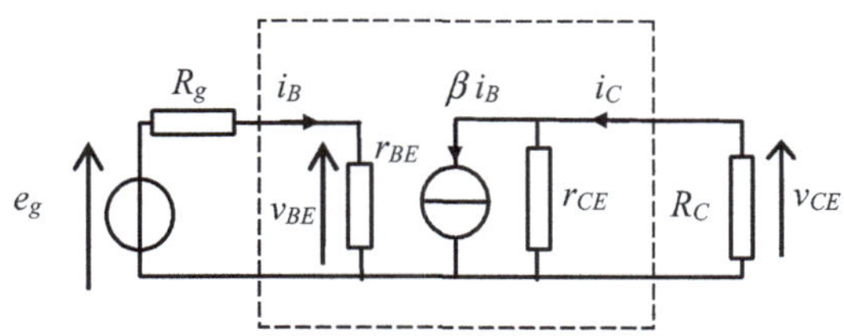

Figure 2.15. *Dynamic circuit of transistor in a common emitter and generator*

We define transistor transconductance $g_m = \dfrac{\beta}{r_{BE}}$ by the ratio of current gain over input resistance. In matrix writing: $\begin{bmatrix} v_{BE} \\ i_C \end{bmatrix} = \begin{bmatrix} r_{BE} & 0 \\ \beta & r_{CE}^{-1} \end{bmatrix} \begin{bmatrix} i_B \\ v_{CE} \end{bmatrix}$, system of hybrid parameters (h) to identify with $\begin{bmatrix} v_1 \\ i_2 \end{bmatrix} = \begin{bmatrix} h_{11E} & h_{12E} \\ h_{21E} & h_{22E} \end{bmatrix} \begin{bmatrix} i_1 \\ v_2 \end{bmatrix}$ or alternatively $\begin{cases} v_1 = h_{11E} i_1 + h_{12E} v_2 \\ i_2 = h_{21E} i_1 + h_{22E} v_2 \end{cases}$ and $g_m = \dfrac{h_{21E}}{h_{11E}}$,

where subscript 1 and 2 for currents and voltages are relative to transistor input and output respectively and where we add subscript *E* for the common emitter circuit.

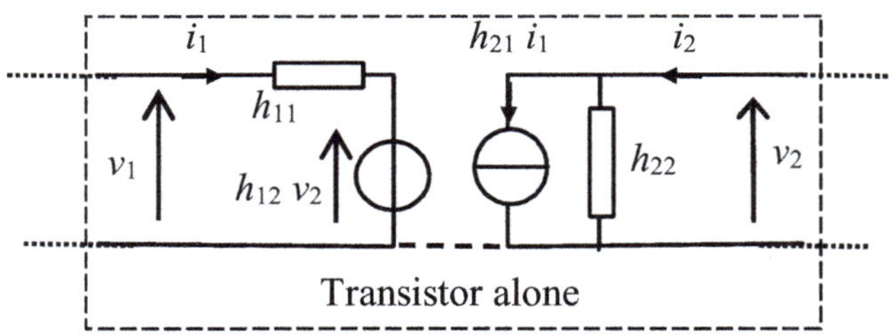

Figure 2.16. *Dynamic circuit with a hybrid two-port network as transistor*

Coefficient h_{11E} is the (dynamic) input resistance: $h_{11E} = \left.\dfrac{v_{BE}}{i_B}\right|_{v_{CE}=0} = r_{BE}$.

Coefficient h_{12E} is zero (no output reaction over input): $h_{12E} = \left.\dfrac{v_{BE}}{v_{CE}}\right|_{i_B=0} = 0$.

Coefficient h_{21E} is the current gain: $h_{21E} = \left.\dfrac{i_C}{i_B}\right|_{v_{CE}=0} = \beta$.

Coefficient h_{22E} is the output conductance: $h_{22E} = \left.\dfrac{i_C}{v_{CE}}\right|_{i_B=0} = r_{CE}^{-1}$.

Accordingly, the common emitter circuit has both a current gain β and a voltage gain $\left.\frac{v_{CE}}{v_{BE}}\right|_{v_{CE}=-R_C i_C} = \frac{-\beta i_B (r_{CE} // R_C)}{r_{BE} i_B} = -g_m (r_{CE} // R_C)$ whose absolute value is significantly greater than when the collector is charged by a sufficiently high-resistance R_C that is close to r_{BE} or larger.

For the common base circuit, we can calculate the hybrid parameters according to those of the common emitter by comparing system $\begin{cases} v_{EB} = h_{11B} i_E + h_{12B} v_{CB} \\ i_C = h_{21B} i_E + h_{22B} v_{CB} \end{cases}$, as shown in Figure 2.17.

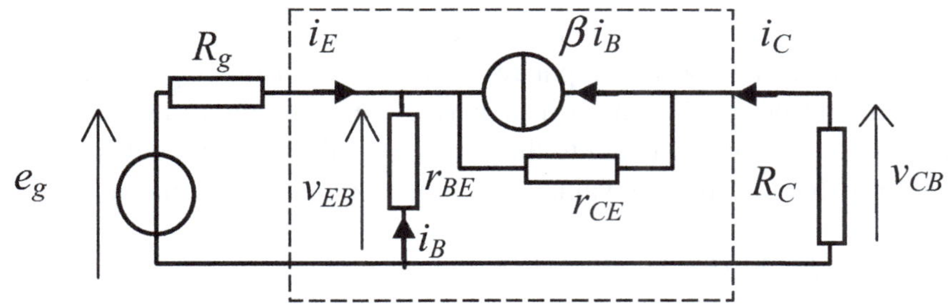

Figure 2.17. *Dynamic circuit of transistor with common base deduced from common emitter parameters*

Coefficient h_{11B} is the (dynamic) emitter resistance:

$$h_{11B} = \left.\frac{v_{EB}}{i_E}\right|_{v_{CB}=0} = \left.\frac{-r_{BE} i_B}{-i_B - \beta i_B - \frac{r_{BE} i_B}{r_{CE}}}\right|_{v_{CB}=0} \approx \frac{r_{BE}}{\beta+1}$$

h_{12B} is the reaction coefficient of output over input:

$$h_{12B} = \left.\frac{v_{EB}}{v_{CB}}\right|_{i_E=0} = \frac{v_{EB}}{v_{EB} + r_{CE}(\beta+1) v_{EB} / r_{BE}} \approx \frac{r_{BE}}{(\beta+1) r_{CE}} < 10^{-4}$$

which can be approximated by zero.

Coefficient h_{21B} is the current gain:

$$h_{21B} = \left.\frac{i_C}{i_E}\right|_{v_{CB}=0} = \left.\frac{\beta i_B + \frac{r_{BE} i_B}{r_{CE}}}{-i_B - \beta i_B - \frac{r_{BE} i_B}{r_{CE}}}\right|_{v_{CB}=0} \approx -\frac{\beta}{\beta+1} \approx -1, \text{ (strictly speaking } -\alpha_F)$$

Coefficient h_{22B} is the output conductance:

$$h_{22B} = \left.\frac{i_C}{v_{CB}}\right|_{i_E=0} = \frac{-i_B}{-r_{BE} i_B - r_{CE}(\beta+1) i_B} \approx \frac{1}{(\beta+1) r_{CE}}$$

Accordingly, the common base circuit has an input resistance (β + 1) times lower than the common emitter circuit and also an output conductance (β + 1) times lower (so an output resistance (β + 1) times greater, which lends it the property of being a near perfect current source) and a current gain close to −1, actually equal to the opposite of the transfer coefficient α_F. We observe that the absolute value of the transconductance is unchanged:

$$|g_m| = \left|\frac{h_{21E}}{h_{11E}}\right| = \left|\frac{h_{21B}}{h_{11B}}\right| = \frac{\beta}{r_{BE}}$$

relative to the common emitter circuit.

Hence, the voltage gain of the circuit reads:

$$\left.\frac{v_{CB}}{v_{EB}}\right|_{v_{CB}=-R_C i_C} = \frac{\beta}{r_{BE}} \left[(\beta+1) r_{CE} \,//\, R_C\right]$$

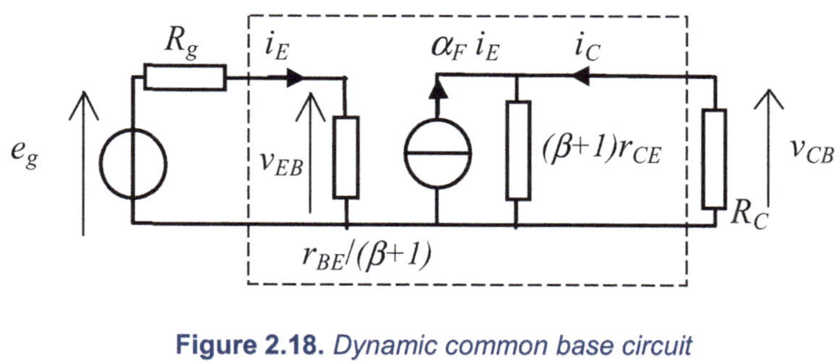

Figure 2.18. *Dynamic common base circuit*

Common collector circuit under dynamic conditions:

$$\begin{cases} v_{BC} = h_{11C} i_B + h_{12C} v_{EC} \\ i_E = h_{21C} i_B + h_{22C} v_{EC} \end{cases}$$

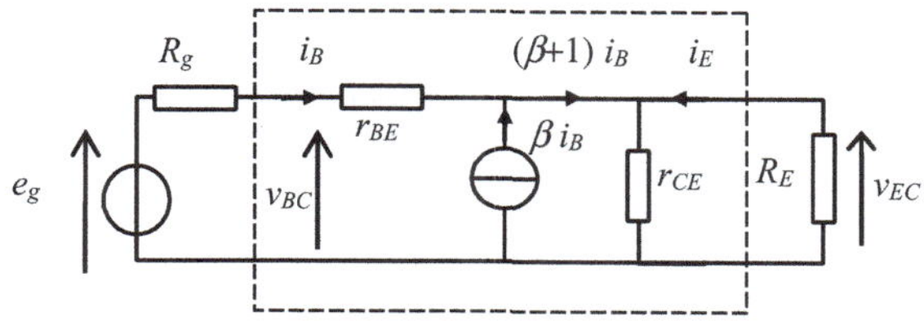

Figure 2.19. *Dynamic common collector circuit deduced from common emitter parameters*

Coefficient h_{11C} is the (dynamic) input resistance:

$$h_{11C} = \left.\frac{v_{BC}}{i_B}\right|_{v_{EC}=0} = \frac{r_{BE} i_B}{i_B} = r_{BE}$$

Coefficient h_{12C} is the reaction coefficient of output over input:

$$h_{12C} = \left.\frac{v_{BC}}{v_{EC}}\right|_{i_B=0} = 1$$

Coefficient h_{21C} is the current gain:

$$h_{21C} = \left.\frac{i_E}{i_B}\right|_{v_{EC}=0} = \frac{-i_B - \beta i_B}{i_B} = -(\beta + 1).$$

Coefficient h_{22C} is the output conductance:

$$h_{22C} = \left.\frac{i_E}{v_{EC}}\right|_{i_B=0} = \frac{i_E}{r_{CE} i_E} = \frac{1}{r_{CE}}$$

However, the properties do not seem particularly clear with this model due to the non-zero output over input reaction coefficient $h_{12C} = 1$. The consequences are more readily interpreted if we calculate external parameters such as input and output resistance in the presence of generator R_g and emitter R_E resistances, together with voltage gain:

$$\left.\frac{v_{BC}}{i_B}\right|_{v_{EC}=-R_E i_E} = \frac{r_{BE} i_B + (\beta+1) i_B (r_{CE} // R_E)}{i_B} = r_{BE} + (\beta+1)(r_{CE} // R_E)$$

$$\left.\frac{v_{EC}}{i_E}\right|_{e_g=0} = \frac{-(R_g + r_{BE}) i_B}{-(\beta+1) i_B - (R_g + r_{BE}) i_B / r_{CE}} \approx \frac{R_g + r_{BE}}{\beta+1} \qquad \text{if} \quad R_g \leq r_{CE}$$

$$\left.\frac{v_{EC}}{v_{BC}}\right|_{v_{EC}=-R_E i_E} = \frac{(\beta+1) i_B (r_{CE} // R_E)}{r_{BE} i_B + (\beta+1) i_B (r_{CE} // R_E)} = \frac{(\beta+1)(r_{CE} // R_E)}{r_{BE} + (\beta+1)(r_{CE} // R_E)} \approx 1$$

The common collector circuit is known as an impedance adaptor since it has an input resistance (β + 1) times higher than the global charge resistance $R_E // r_{CE}$, an output resistance (β + 1) times lower than the global source resistance r_{BE} + R_g, a current gain (β + 1) in absolute value and a voltage gain very slightly lower than 1. So, it disturbs the preceding stage only very slightly due to its high input impedance and acts as a near perfect voltage source for the following stage.

2.4.2. *Small-signal high-frequency equivalent circuit*

Under conductive regimes, we must add the capacitances of the depletion zone $C_{B'C}$ for the CB' junction and the diffusion capacitance $C_{B'E}$ for the B'E junction under forward bias, together with that of the access resistance $r_{BB'}$ to the effective base B'. Hence, the Giacoletto circuit for the bipolar transistor in common emitter is shown in Figure 2.20.

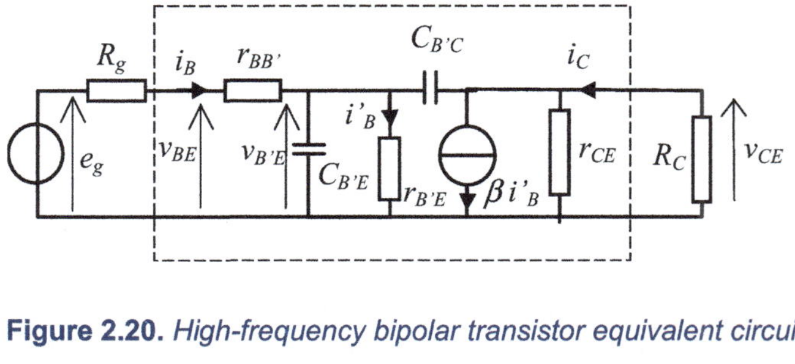

Figure 2.20. *High-frequency bipolar transistor equivalent circuit*

Current $\beta i'_B$ can be replaced by $g_m v_{B'E}$ without disadvantage.

The active base current is that which runs through the resistance of the base emitter junction, that is $r_{B'E}$. This entails the introduction of a cut-off frequency in the current gain under sinusoidal regimes that is $\dfrac{1}{2\pi r_{B'E}(C_{B'E}+C_{B'C})}$ (output in short-circuit), just below $f_\beta = \dfrac{1}{2\pi r_{B'E} C_{B'E}}$. The cut-off frequency at $R_C \neq 0$ is still lower since the negative voltage gain amplifies the current in $C_{B'C}$, which then appears from the base as a capacitance $(1 + A_v)\, C_{B'C}$ if A_v is the absolute value of the voltage gain under low frequency (Miller effect, see exercises).

In the common base circuit, the input resistance is divided by $(\beta + 1)$, so $r_{B'E}$ as well, which involves a time constant $(\beta + 1)$ times lower and consequently a current gain cut-off frequency (close to 1 with common base) that is $(\beta + 1)$ times higher: $f_T = (\beta+1) f_\beta = \dfrac{\beta+1}{2\pi r_{B'E} C_{B'E}}$, a value often given in provider documentations, called transition frequency. Capacitance $C_{B'C}$ occurs in parallel on output and the Miller effect then disappears.

In any case, $r_{B'E}\, C_{B'E}$ is equal to τ, the lifetime of minority carriers in the base, and is not markedly subject to the operating point, so f_T and f_β too. Consequently, this circuit can also be suited for simulating operation under strong signals.

2.5. Power amplification: classes of amplification

The common emitter circuit described and studied above is a class A amplifier, which means that the operating point (or rest point, that is at $e_g = 0$) of the transistor corresponds to static non-zero currents and (bias or rest) voltages, equal to the mean value of total currents and voltages.

The yield in sinusoidal regimes, that is, the power ratio in charge resistance R_C over the power provided by supply E is then limited to 25% in the best cases (rest point in the middle of the line segment of the load equation, here $I_{C1} = E/2R_C$):

$$\frac{R_C\left(E/2\sqrt{2}R_C\right)^2}{E^2/2R_C} = \frac{1}{4}$$

In order to increase yield, the rest point is best positioned at zero current, with the use of two common collector transistors for each current direction in the load with the class B circuit.

When $e_g > U_0$, only I_{c1} is $\neq 0$, and when $e_g < -U_0$, only $I_{c2} \neq 0$.

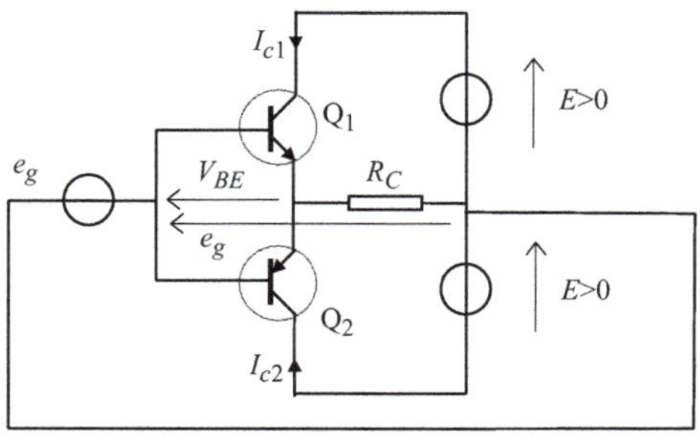

Figure 2.21. *Class B circuit*

In sinusoidal regime, if $e_{gmax} = E >> U_0$, the E sources provide each a current with an average value $E/\pi R_C$ and the power in R_C is $\frac{\left(E/\sqrt{2}\right)^2}{R_C}$ for a maximum yield of $\frac{\left(E/\sqrt{2}\right)^2/R_C}{2E^2/\pi R_C} = \frac{\pi}{4} = 78\%$.

The distortion due to the cancellation of currents when $-U_0 < e_g < U_0$ can be eliminated by biasing the BE junctions under a $2U_0$ voltage, so that each transistor remains slightly conductive when the other provides the load current. In order to ensure thermal stability, it is best to add emitter resistances. A low biasing current permanently crosses the two transistors, somewhat deteriorating the yield of this amplification circuit under class AB.

In case of a load having a resonant impedance (parallel $R\ L\ C$ circuit), we can use voltage pulses to control transistor conduction, with a rest point at zero current (C class amplification) even in the common emitter circuit. If the pulse frequency is equal to the resonance frequency, only sinusoidal voltage at the fundamental frequency is amplified and we can then obtain yields approaching unity (see exercise).

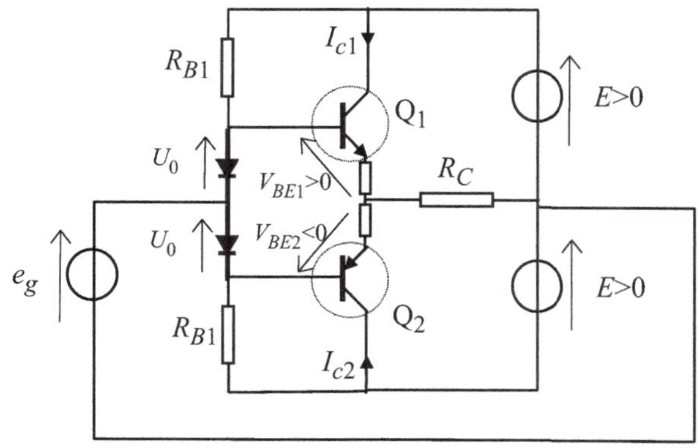

Figure 2.22. *Class AB circuit*

2.6. Bipolar transistor switching on resistive loads

The charging and discharging phenomena happening in capacitances and transfer of minority carriers stored in the EB diode are the same as those described in the pn diode (cf. chapter 1 on pn diode). Furthermore, we must account for the change in operating regime of the CB diode moving from reverse bias to forward bias when the collector current reaches I_{Csat}, a value that puts the transistor in saturation. The common emitter circuit is the only one presently in use, since the common base circuit requires too much control current and the common collector too much control voltage. In this case (Figure 2.14), $I_{Csat} = (E - V_{CEsat}) / R_C$ is still set by the external circuit (as V_{CEsat} = 0.05V to 0.2V typically, much smaller than the power supply voltage E) while $I_C = \beta I_B$ no longer applies. We write $I_{Bmin} = I_{Csat} / \beta$; this is the base current for which transistor switching from the conductive state to the saturation state occurs, passing through $V_{CB} = 0$. However, the time t_r taken to reach this current I_{Csat} is set by the final current I_{B2} through the conductive state where relation $I_C = \beta\ I_B$ still applies, as shown in Figure 2.23.

On opening, the storage t_s and fall times t_f are set by the evolution of $I_C\ (t)$, which is governed by $i_B(t) = \dfrac{dQ_s}{dt} + \dfrac{Q_s(t)}{\tau}$, the initial ($I_{B2}$) and final ($-I_{B1}$) regimes, and the only part of the base current that participates in the transistor effect, that is $\dfrac{Q_s(t)}{\tau}$, the first term being a displacement current.

Switching on closure is accelerated by increasing I_{B2} (forced saturation) but switching on opening is then slowed. A first solution consists of increasing I_{B1}, which accelerates opening. Another solution consists of preventing the base current from actually reaching I_{B2} after time t_r by placing a low threshold (≈ 0.3 V), anti-saturation Schottky diode between base and collector (dotted in Figure 2.23). When the collector potential falls below that of the base, the excess base current is shunted into the collector by the diode, which avoids storing more minority charges than required to saturate the transistor. This method is used systematically in the logical circuits of the type TTL Schottky, hence their name (S, LS, ALS series, etc.).

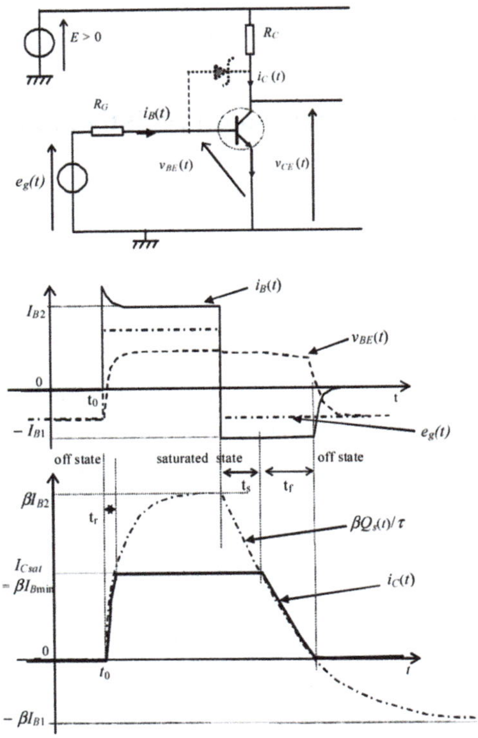

Figure 2.23. *Circuit and signals in the common emitter transistor switching a resistive load (with the hypothetical collector current in mixed line when extrapolated in the linear regime)*

2.7. Components based on the pnpn structure

Tripole type components are intended to control significant power levels, higher than several dozen watts and up to several kilowatts. The use of such components touches on the domain of power electronics and as a consequence falls outside of this book's remit. However, since part of their operation is based on that of the bipolar transistor, and since the pnpn structure is also the origin of the isolated gate bipolar transistor (IGBT) device that has become so important and is studied in chapter 3, their development and main properties are described in the following.

2.7.1. *pnpn diode (or Shockley diode)*

This component comprises four alternate doping zones that we will refer to as P1, N1, P2 and N2, as shown in Figure 2.24. In a logical manner, P1 is also anode A while N2 is the cathode, called K. In reverse bias, that is, when V_{AK} is negative, both junctions P1-N1 and P2-N2 are in reverse bias while N1-P2 is in forward bias and consequently tolerates a negligible drop in voltage. Since zone N1 is the most weakly doped, it takes the greater part of the voltage applied between P1 and N2 while the P2-N2 junction takes a lower voltage but an electric field that can be higher due to the more significant doping concentration in these zones. In this case, the current crossing the diode is negligible due to the two reverse biased junctions and remains such so long as these P1-N1 or P2-N2 junctions do not reach avalanche regime.

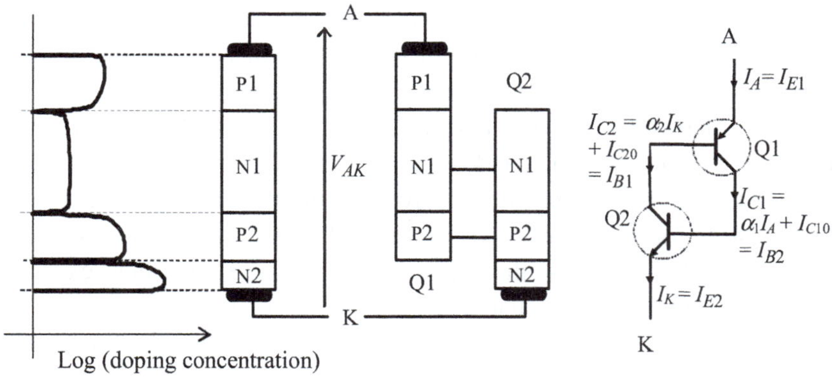

Figure 2.24. *Doping concentration and electric circuitry of the pnpn diode*

In case of a positive V_{AK} voltage, only the central junction N1-P2 is in reverse bias while P1-N1 and P2-N2 junctions undergo forward bias. It follows that from this state the diode is crossed by a negligible current, as we see elsewhere for any reverse bias status. However, by means of a cooperative effect, it is possible to give rise to a current that will be amplified to the point of making all the structure conductive, since P1 and N2 can act as emitters in transistors P1-N1-P2 and N2-P2-N1 respectively, in accordance with the circuitry represented in Figure 2.24. The directions have been altered so that all of the currents are positive, while we have renamed transistors P1-N1-P2 and N2-P2-N1 as Q1 and Q2 respectively, with the common base current gains α_1 and α_2 and the leakage currents from the collector I_{C10} and I_{C20}.

In addition to the relations between emitter currents and collector currents noted in the figure, there is also a relation between the base current of Q1 and I_{C2} : $(1 - \alpha_1) I_A - I_{C10} = I_{C2}$ and between the base current of Q2 and I_{C1}: $(1 - \alpha_2) I_K - I_{C20} = I_{C1}$. The current conservation in Q2 also results in $I_K = I_{C2} + I_{C1}$ and $I_K = I_A$ since it is a diode, that is, a bipolar device. By accounting for these relations, we find:

$$I_A = (1-\alpha_1)I_A - I_{C10} + (1-\alpha_2)I_A - I_{C20} \text{ or, alternatively,}$$

$$I_A = \frac{I_{C10} + I_{C20}}{1-(\alpha_1 + \alpha_2)}$$

For the Q2 transistor, the decreasing progression of emitter N2 doping impurities to collector N1 is necessary for the transistor effect to be significant, which results in α_2 being quite close to 1, even if we cannot decrease the thickness of base P2 so as to not compromise the withstand voltage of the P2-N2 junction in reverse bias. This is not the case for α_1 which is substantially lower than 1. However, the increase in V_{AK} produces two effects that together act to increase I_A. On the one hand, since these coefficients α_1 and α_2 increase when the bases' neutral zones see a decrease in their thickness under the effect of the increased total voltage V_{AK} (see section 2.1), the total $\alpha_1+\alpha_2$ manages to appraoch 1, which makes the denominator of the previous expression approach zero. On the other hand, the numerator of the expression of I_A is increased due to the generation of electron-hole pairs, favored by the increased electric field and the extension of the space charge zone from the N1-P2 junction. At a certain voltage $V_{AK} = V_{trig}$, due to the varying numerator and denominator, current I_A exceeds critical value I_{trig} then increases abruptly through a cumulative effect due to the close-looped circuit of the two transistors, resulting in a massive injection of carriers in both central zones. This has the effect of considerably increasing the conductivity of the neutral part of these zones, and moreover the two transistors go from off state to saturation due to the overwhelming increase in the concentration of carriers in the depleted zone of the central N1-P2 junction. This phenomenon entails the near cancellation of the voltage difference in this zone, which corresponds to the collector–base junction both for Q1 and for Q2, and which had previously prevented the current flow. Here, the device presents a negative differential resistance (see Figure 2.25) if the forward voltage V_{AK} exceeds a certain value V_{trig}, typically somewhere between several dozen and several hundred volts, and simultaneously if I_A exceeds I_{trig}. After triggering, the junctions remain in what is called a "latch up" state, where both transistors are saturated so long as current does not fall below a certain threshold I_{hold}, called the hold current, of the same order of magnitude as I_{trig} but slightly lower. The triggering voltage can be decreased if V_{AK} increases quickly as it results in a displacement current in the initially non-conductive N1-P2 junction, equivalent to a capacitance, and consequently through the rest of the diode. Since this current is added to the static leak current, it can reach the triggering current I_{trig} for a lower V_{AK} voltage than in the static regime. Due to the presence of a junction in the saturation regime, draining the stored minority charges out of the BE junctions takes a certain time, typically in the order of 100–200 μs, on return to the off state. This switching time limits the use of this type of device to frequencies below kilohertz, and more specifically to that of the electric grid.

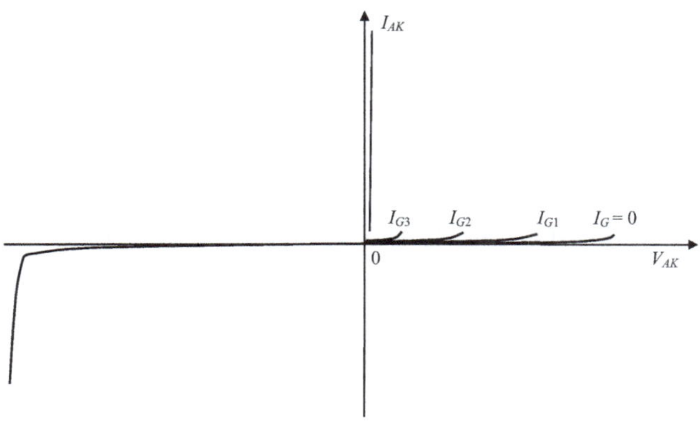

Figure 2.25. *Current voltage characteristics of the pnpn diode (I_G = 0) and of the thyristor for rising gate currents ($I_{G3} > I_{G2} > I_{G1}$)*

2.7.2. Controlled rectifier or thyristor

In order to control pnpn diode triggering, we need only add a rising gate current I_G on the base of npn transistor Q2 (Figure 2.26), which has the effect of increasing the numerator of the previous expression of I_A and thereby allowing for the critical regime to be reached at a lower V_{AK} voltage (see Figure 2.25). Indeed, the modified pnpn diode equations are written as: $I_K = I_{C2} + I_{C1} + I_G$; $I_K = I_A + I_G$ and $(1 - \alpha_2)\, I_K - I_{C20} = I_{C1} + I_G$. Hence, it follows that: $I_A = I_{C2} + I_{C1} = (1 - \alpha_1)\, I_A - I_{C10} + (1 - \alpha_2)\,(I_A + I_G) - I_{C20} - I_G$; or alternatively:

$$I_A = \frac{\alpha_2 I_G + I_{C10} + I_{C20}}{1-(\alpha_1+\alpha_2)}$$

The higher the gate current in the numerator, the less necessary it is for the denominator to approach zero and for voltage V_{AK} to be large in order to reach the triggering current I_{trig}. This gate current effect is still more effective in that it can very easily be several orders of magnitude higher than the total of the pnpn structure leakage currents. It is obtained by polarizing the base emitter junction of Q2 (P2-N2) in the direction of flow, typically in a pulsed regime.

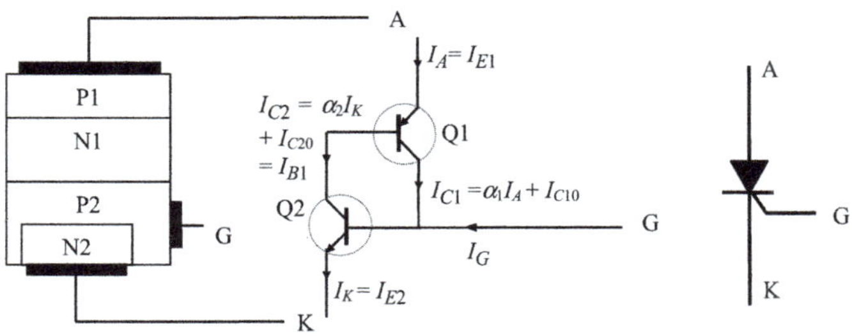

Figure 2.26. *Thyristor circuit and its symbol*

This controlled rectifier is useful in adjusting power delivered to a load, with the device being supplied by alternating voltage typically derived directly from the grid or through a transformer, since current extinction occurs automatically when the current becomes lower than the holding current. It can be triggered under forward voltage either approaching the passage of voltage through zero during several periods (controlled by wave train) or on each forward half-wave at a set instant and after the passage of voltage through zero (controlled by the phase angle), allowing for an adjustable half-wave rectification. This type of component can tolerate reverse voltages approaching the kilovolt or slightly higher and currents of several hundred amperes.

Double-wave rectification requires two thyristors set end-to-end and suitable control circuits for both gate currents. An alternative solution consists of using the symmetrical components described in the following.

In order to control not only current triggering but also its extinction, we devise a structure derived from the pnpn diode by adding a well identical to that of the gate zone, albeit more strongly doped (p+ in Figure 2.27). Triggering occurs in the same manner as for the thyristor, by a positive V_{GK} impulse. Extinction is obtained by derivating a part of the current at the P2-N2 junction into the G gate electrode by means of a negative V_{GK} impulse capable of starting an avalanche effect made possible by the strong doping in this part of the gate. The Q2 transistor can then be desaturated and the increase in voltage difference at the transistor's terminals leads to a still greater decrease in current that can quickly reach the extinction value. This component, called the gate turned-off (GTO) thyristor or simply GTO, tends to be replaced by the IGBT, which is more flexible and more reliable in terms of triggering, as we will study in Chapter 3.

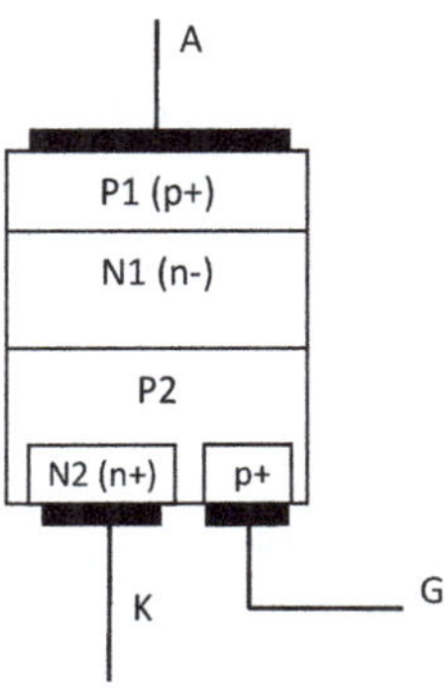

Figure 2.27. *Structure of GTO thyristor (gate turned-off thyristor)*

2.7.3. *Diacs and triacs*

Parallel (or end-to-end) installation of the components above can be done in the same semiconductor volume to obtain symmetrical current–voltage characteristics around the origin. In this way, the diac (Figure 2.28) has symmetrical characteristic with a negative conductance range for each polarity. It is assembled like a symmetrical bipolar transistor without a base electrode. Under the effect of sufficient applied voltage, typically in the order of 30V, one of these diodes goes into the avalanche regime and current is injected into the other diode that acts as the base emitter junction of a bipolar transistor, in turn injecting the charge carriers into the junction that was previously in the avalanche regime. Due to the transistor effect, the latter then acts as a base collector junction in the saturated state, provoking a drop in total voltage and negative conductance. Diacs are generally used in series with the gate of a thyristor or a triac for triggering control.

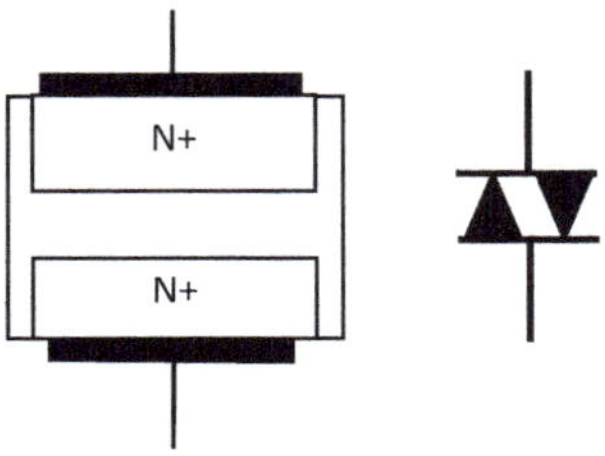

Figure 2.28. *Structure of diac and its symbol*

The triac (Figure 2.29) comprises two parallel structures, one PNPN and the other NPNP, between the two electrode terminals M2 and M1. The gate comprises two parallel wells, one of type n and the other of type p, in the p-type zone connected to M1. In this way, four triggering regimes are possible, two for positive $V_{M2\text{-}M1}$ and two for negative $V_{M2\text{-}M1}$. For both of these $V_{M2\text{-}M1}$ polarities, the gate current can be either positive or negative.

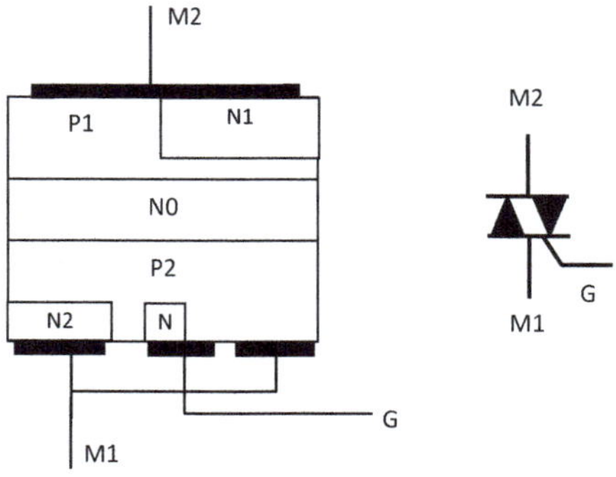

Figure 2.29. *Structure of a triac and its symbol*

Each of these four regimes (or quadrant operations) corresponds to one association of two Q1 and Q2 transistors, loop connected as in the thyristor. In any case, M1 and M2 are the emitters in both transistors.

Quadrant	$V_{M2\text{-}M1}$	$V_{G\text{-}M1}$
1	>0	>0
2	>0	<0
3	<0	<0
4	<0	>0

EXERCISE.– Draw the four circuits with Q1, Q2 and the real current directions.

Quadrants 1 and 3 require a lower triggering current than the others and are more widely used, with the exception of low power triacs controlled by a logical circuit that only provides one polarity for $V_{G\text{-}M1}$. One type of triac called "alternistor" only works in the first three quadrants. All of these components require circuits for control and to help switching, particularly with inductive or capacitive loads, which is subject to discussion in power electronics and so falls outside of our remit here. Their gate control, performed by optoelectronic components described briefly in section 2.8, is of significant value in resolving insulation issues.

2.8. Phototransistors and optically controlled components

As indicated in chapter 1, we can create electron-hole pairs in a junction using the light emitted by an electroluminescent diode made with a III-V semiconductor whose forbidden bandgap is greater than that of the silicon with which the junction was made. This is what is achieved in a phototransistor, whose geometry (Figure 2.30) allows for the illumination of the collector–base junction, which is reverse biased and consequently benefits from the greater conversion yield of photons into mobile charge carriers. Since phototransistors only have two terminal electrodes, one for the emitter and the other for the collector, the current also crosses the base emitter junction, which is consequently forward biased since the two junctions are end-to-end. Accordingly, a transistor effect can occur, allowing for greater sensitivity than that of a simple opto-coupler composed of an electroluminescent diode and a photodiode. However, avalanche photodiodes, which incorporate a very slightly doped zone in between the n and p zones, are also able to take advantage of a current gain due to their optimized design allowing permanent avalanche.

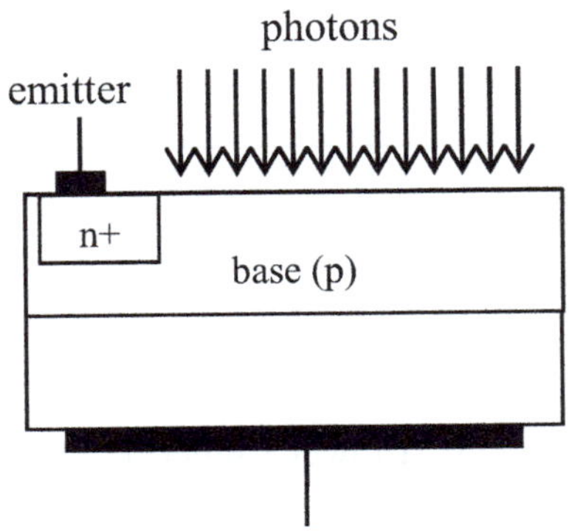

Figure 2.30. *Phototransistor npn*

Multithyristors assemblies and triacs can be controlled advantageously by one or two phototransistors, which in some components are integrated in a single case, hence the name photothyristor or phototriac. This type of association can be most beneficial since it allows for insulation of the gate control circuit vis-à-vis the thyristor or triac power circuit.

2.9. Exercises

2.9.1. *Class A amplification*

The transistor current gain is constant and equal to β. Voltage v_e is alternative and the binding capacitance C_L has negligible impedance in the alternative regime.

– Static regime ($v_e = 0$):

Calculate current $I_B = I_{B1}$ according to R_B, V_{BE}, E. Deduce $I_C = I_{C1}$ according to R_B, V_{BE1}, E and β. Calculate $V_{CE} = V_{CE1}$ and establish what must be verified so that V_{CE1} = E/2, that is, so that the operating point is in the middle of the load line (draw the graph of I_C (V_{CE}) and the load line). We can consider $E >> V_{BE1}$.

– Operation as amplifier in common emitter circuit. Nonlinearity with respect to R_G:

We must replace resistance R_B by a source of constant, ideal current I_{B1}. We write $V_{BE} = V_{BE1} + v_{BE}$, $V_{CE} = V_{CE1} + v_{CE}$, $I_{CE} = I_{C1} + i_C$, etc. where the lowercase variables represent the alternative variation of voltages or currents. The base current in forward bias is given by $I_{B0}\exp\left(\frac{V_{BE}}{\eta V_T}\right)$ and that $I_{B1} = I_{B0}\exp\left(\frac{V_{BE1}}{\eta V_T}\right)$ with $\eta V_T \approx 30$ mV at ambient temperature.

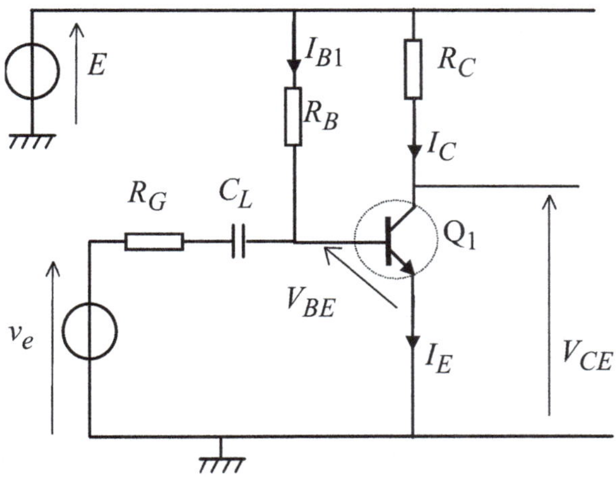

Find the expression of the total base current (I_{B1} static + alternative variation i_B) using a series expansion. Show that the first term of i_B is equal to v_{BE} / r_{BE}, where r_{BE} represents the dynamic resistance of the base–emitter junction, and that the relation between v_{BE} and v_e is typically nonlinear. What condition on v_{BE} will let us improve linearity?

In both boundary cases $R_G = 0$ (approached by an ideal voltage source) and $R_G >> r_{BE}$ (approached by an ideal current source), if v_e is the sinusoidal alternative, which variations remain sinusoidal and which do not (distortion)? In both cases, deduce the expressions for transistor voltage gain $\frac{v_{CE}}{v_{BE}}$ and circuit voltage gain $\frac{v_{CE}}{v_e}$ disregarding the terms of degree greater than or equal to 2 ("small signals" approximation). Conclusion? Establish the equivalent small signal circuit for the only alternative variations according to the equations of the circuit reduced to only linear terms and after simplification of static terms.

Answer:

$$I_{B1} = \frac{E - V_{BE1}}{R_B} \text{ hence } I_{C1} = \beta \frac{E - V_{BE1}}{R_B}$$

$V_{CE1} = E - R_C I_{C1}$ and, carrying forward the expression of I_{C1}, we have:

$$V_{CE1} = E\left[1 - \beta\frac{R_C}{R_B}\right] + \beta\frac{R_C}{R_B}V_{BE1}$$

If V_{CE1} = $E/2$: $\frac{E}{2} = \beta\frac{R_C}{R_B}[E - V_{BE1}]$ and as $E >> V_{BE1}$, we obtain $\frac{R_B}{R_C} = 2\beta$.

We have: $I_B = I_{B1} + \frac{v_e}{R_G + r_{BE}} = I_{B0}\exp\left(\frac{V_{BE1} + v_{BE}}{\eta V_T}\right)$.

Through series expansion of the exponential of $v_{BE}/\eta V_T$, we obtain:

$$I_B = I_{B1} + \frac{v_e}{R_G + r_{BE}} = I_{B0}\exp\left(\frac{V_{BE1}}{\eta V_T}\right)\left[1 + \frac{v_{BE}}{\eta V_T} + \frac{1}{2!}\left(\frac{v_{BE}}{\eta V_T}\right)^2 + \frac{1}{3!}\left(\frac{v_{BE}}{\eta V_T}\right)^3 + \cdots\right]$$

with $i_B = \frac{v_e}{R_G + r_{BE}}$ and $I_{B1} = I_{B0}\exp\left(\frac{V_{BE1}}{\eta V_T}\right)$; simplifying by I_{B1}, we may deduce:

$i_B = I_{B1}\frac{v_{BE}}{\eta V_T} + I_{B1}\left[\frac{1}{2!}\left(\frac{v_{BE}}{\eta V_T}\right)^2 + \frac{1}{3!}\left(\frac{v_{BE}}{\eta V_T}\right)^3 + \cdots\right]$, in which the first term is effectively equal to $\frac{v_{BE}}{r_{BE}}$ since the denominator $\frac{\eta V_T}{I_{B1}}$ is indeed equal to the dynamic resistance of the base emitter junction that we can calculate directly from the differential of $I_{B1} = I_{B0} \exp\left(\frac{V_{BE1}}{\eta V_T}\right)$.

Since $v_{BE} = v_e - R_G\, i_B$, so $v_{BE} = v_e - R_G\left[I_{B1}\frac{v_{BE}}{\eta V_T} + \frac{I_{B1}}{2!}\left(\frac{v_{BE}}{\eta V_T}\right)^2 + \frac{I_{B1}}{3!}\left(\frac{v_{BE}}{\eta V_T}\right)^3 + \cdots \right]$.

Or alternatively: $v_{BE} = v_e - \frac{R_G}{r_{BE}} v_{BE}\left[1 + \frac{1}{2!}\frac{v_{BE}}{\eta V_T} + \frac{1}{3!}\left(\frac{v_{BE}}{\eta V_T}\right)^2 + \cdots \right]$.

If $R_G \to 0$, then $v_{BE} \to v_e$. Voltage v_{BE} is sinusoidal if v_e is sinusoidal and i_B is not since it was calculated in the brackets of the series expansion in the penultimate expression. So, the only solution letting us improve the linearity of i_B is that the terms of order ≥ 2 are low relative to 1, which is to say that $v_{BE} << \eta V_T$, so far below 30 mV.

If $R_G >> r_{BE}$, we have $v_{BE} << v_e$ since:

$$v_{BE} = \frac{v_e}{1 + \frac{R_G}{r_{BE}}\left[1 + \frac{1}{2!}\frac{v_{BE}}{\eta V_T} + \frac{1}{3!}\left(\frac{v_{BE}}{\eta V_T}\right)^2 + \cdots \right]}$$

has a denominator >> 1. We have the equivalent of a current source that can supply the BE junction since the most significant voltage fall will occur at the terminals of R_G and not r_{BE}; current i_B is consequently $\approx v_e / R_G$ and is sinusoidal if v_e is sinusoidal. Since we still have $i_B = I_{B1}\frac{v_{BE}}{\eta V_T} + \frac{I_{B1}}{2!}\left(\frac{v_{BE}}{\eta V_T}\right)^2 + \frac{I_{B1}}{3!}\left(\frac{v_{BE}}{\eta V_T}\right)^3 + \cdots$ and i_B is sinusoidal, v_{BE} cannot be unless the terms of the order ≥ 2 are low relative to 1, that is if $v_{BE} << \eta V_T$ = 30 mV.

In both cases, we must calculate the expressions of transistor voltage $\frac{v_{CE}}{v_{BE}}$ and circuit gains $\frac{v_{CE}}{v_e}$ by disregarding the terms of degree greater than or equal to 2 ("small signals" approximation) from which we have the variation of the collector current in the load resistance: $v_{CE} = -\beta i_B R_C = -\beta \frac{v_{BE}}{r_{BE}} R_C$ or alternatively$= -\beta \frac{v_e}{r_{BE}} R_C$. Indeed, $V_{CE} = V_{CE1} + v_{CE}$; $I_C = I_{C1} + i_C$ and $V_{CE} = E - R_C I_C$; so $V_{CE1} + v_{CE} = E - R_C I_{C1} - R_C i_C \Rightarrow v_{CE} = -R_C i_C$.

When $R_G = 0$, both gains are identical, being $-\beta \frac{R_C}{r_{BE}}$.

When $R_G >> r_{BE}$, transistor gain remains $-\beta \frac{R_C}{r_{BE}}$ but that of the circuit becomes $\frac{v_{CE}}{v_e} = -\beta \frac{R_C}{R_G}$ whose absolute value is much smaller than $\beta \frac{R_C}{r_{BE}}$.

The equivalent small signals circuit assembly for the only alternative variation is reduced to:

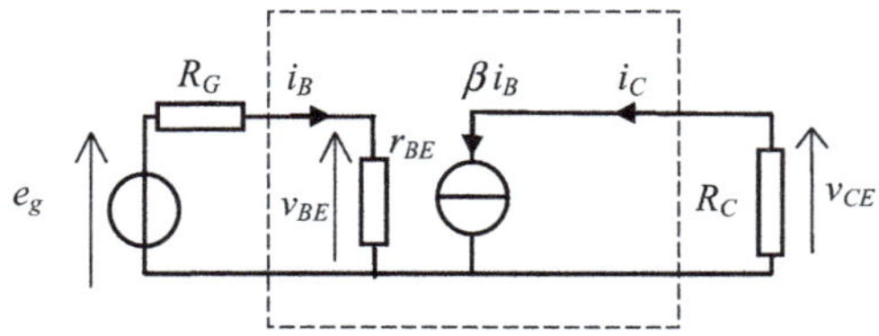

2.9.2. 4 Quadrant multiplier

i) Differential stage

By breaking down each V_{BE} voltage according to the differential voltage $V_D = V_{BE1} - V_{BE2}$ and the common mode voltage $V_{MC} = \frac{V_{BE1} + V_{BE2}}{2}$ and assuming that for both transistors the collector current is in the form of $I_c = I_{c0}.\exp\left(\frac{V_{BE}}{V_T}\right)$, show that we have:

$$I_{c1} = I_1 + I_1.\tanh\left(\frac{V_D}{2V_T}\right) \text{ and } I_{c2} = I_1 - I_1.\tanh\left(\frac{V_D}{2V_T}\right)$$

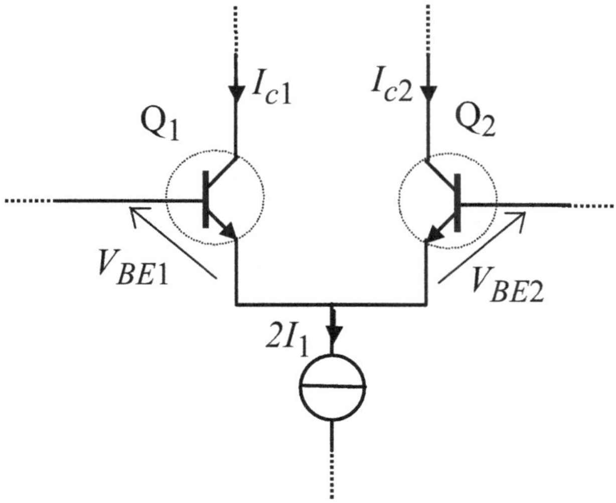

Determine the transconductance around point $V_D = 0$.

Plot both currents according to V_D.

ii) Linearized differential stage

By using the linearized model for BE diodes assumed identical, show that if R_X is very high relative to dynamic resistances r_{BE} then:

$(\approx \frac{V_T}{I_E}$, where I_E is the emitter current $\approx I_C$),

$V_X \approx R_X I_X$ so $I_{c1} = I_1 + \frac{V_X}{R_X}$ and $I_{c2} = I_1 - \frac{V_X}{R_X}$.

iii) 4 Quadrant multiplier

We will disregard the base currents relative to the emitter and collector. Only "T" junctions represent a connection.

Determine relations $I_{C1}(V_Y)$, $I_{C2}(V_Y)$, $I_{C3}(V_X)$, $I_{C4}(V_X)$.

Determine differences $I_{C1} - I_{C2}$, $I_{C6} - I_{C5}$, $I_{C7} - I_{C8}$ for pairs Q_9Q_{10}, Q_5Q_6, Q_7Q_8 respectively, according to V_D.

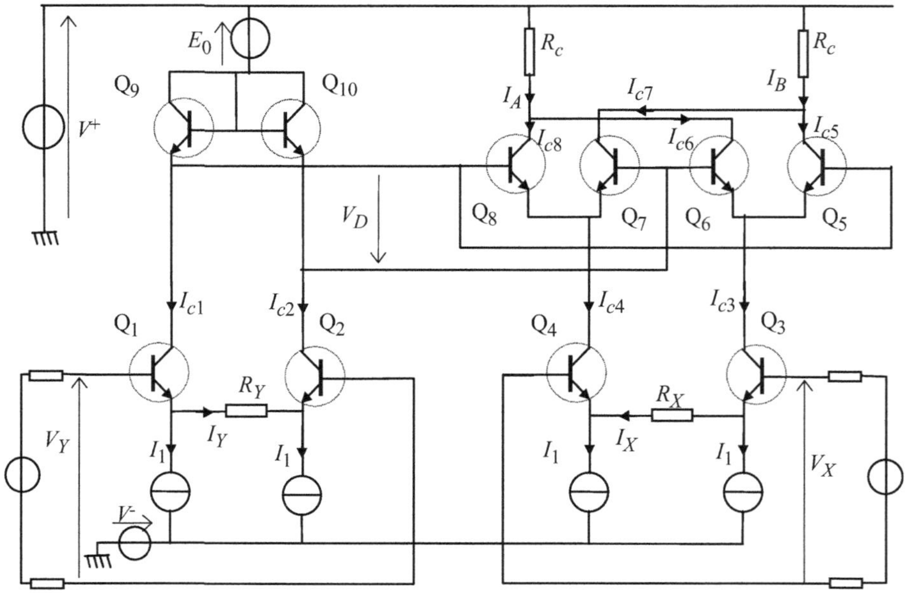

From this, deduce $I_A - I_B$ according to V_Y, V_X, R_X, R_Y, I_1.

Application to signal product: decompose difference $I_A - I_B$ into purely sinusoidal signals when $V_X = V_1 \cos(2\pi f_X t)$; $V_Y = V_1 \cos(2\pi f_Y t)$.

Give the aspect of $I_A - I_B$ when (a) $f_X >> f_Y$ using the signal in the form of a product; (b) $f_X + f_Y >> |f_X - f_Y|$ using the signal decomposed into a sum.

Answer:

i) Differential stage

$$V_{BE1} = V_{MC} + \frac{V_D}{2} \text{ and } V_{BE2} = V_{MC} - \frac{V_D}{2}$$

Hence, $I_{c1} = I_{c0}.\exp\left(\frac{V_{MC}}{V_T}\right)\exp\left(\frac{V_D}{2V_T}\right)$ and $I_{c2} = I_{c0}.\exp\left(\frac{V_{MC}}{V_T}\right)\exp\left(-\frac{V_D}{2V_T}\right)$.

Since $I_{c1} + I_{c2} = 2I_1$, $I_1 = I_{c0}.\exp\left(\frac{V_{MC}}{V_T}\right)\cosh\left(\frac{V_D}{2V_T}\right)$ and $I_{c1} - I_{c2} = 2I_{c0}.\exp\left(\frac{V_{MC}}{V_T}\right)\sinh\left(\frac{V_D}{2V_T}\right) = 2I_1\tanh\left(\frac{V_D}{2V_T}\right)$

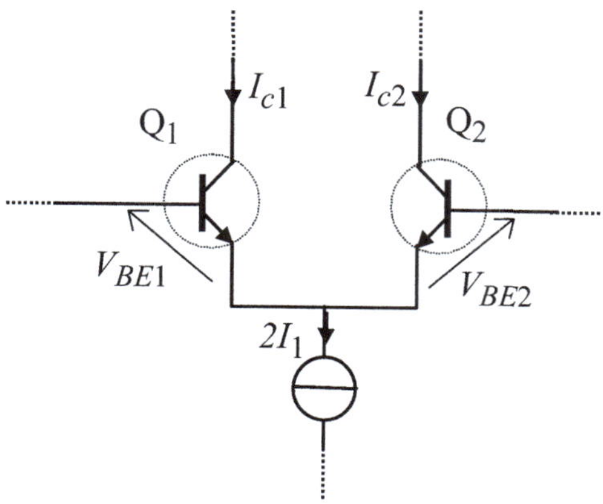

By combining the total and the difference of the two currents, we obtain: $I_{c1} = I_1 + I_1.\tanh\left(\frac{V_D}{2V_T}\right)$ and $I_{c2} = I_1 - I_1.\tanh\left(\frac{V_D}{2V_T}\right)$.

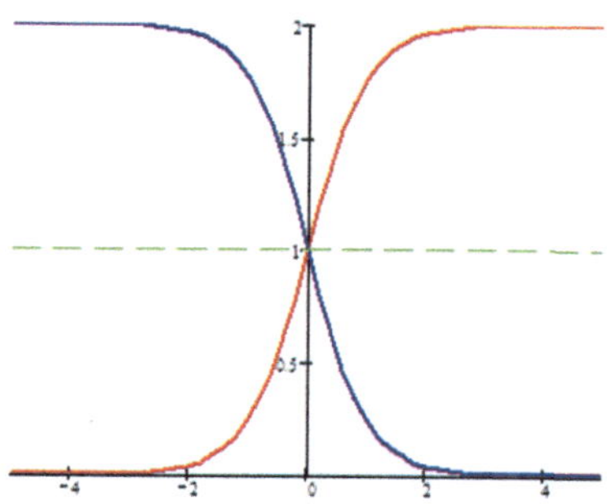

Normalized currents I_{C1}/I_1 & I_{C2}/I_1 according to $V_D/2V_T$

Around point $V_D = 0$, $dI_{c1} = I_1\frac{dV_D}{2V_T}$, which determines a transconductance $\frac{I_1}{2V_T}$.

ii) Linearized differential stage

Voltage drop V_X between bases is made up of $r_{BE}(i_{B1} - i_{B2}) + R_X I_X$ since both threshold voltages compensate for each other. Since I_X is a differential collector current if base currents are neglected, and is so far greater than a differential base current, and since R_X is high relative to dynamic resistances r_{BE}, the first term is negligible and so $V_X \approx R_X I_X$.

So $I_{c1} = I_1 + \frac{V_X}{R_X}$ and $I_{c2} = I_1 - \frac{V_X}{R_X}$

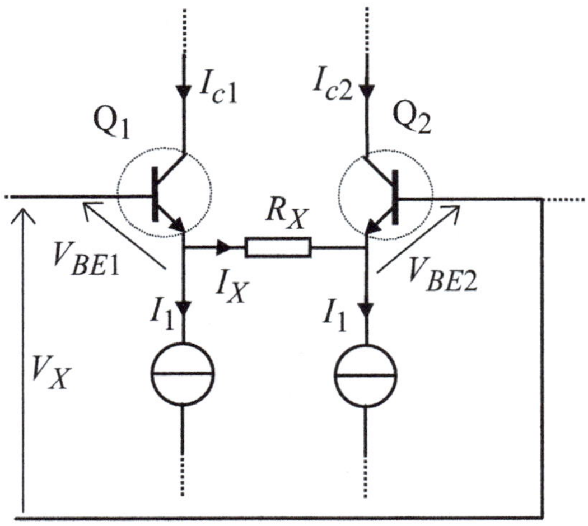

iii) 4 Quadrant multiplier

We will disregard the base currents relative to the emitter and collector. Only "T" junctions represent a connection.

For Q_9-Q_{10}: $V_D = V_{BE9} - V_{BE10}$ and as the emitter currents are considered to be identical to those of the collector, we have:

$$\begin{cases} I_{C1} - I_{C2} = 2I_1 \tanh\left(\dfrac{V_D}{2V_T}\right) \\ I_{C1} + I_{C2} = 2I_1 \end{cases} \text{ and for } Q_1\text{-}Q_2 \begin{cases} I_{C1} - I_{C2} = 2\dfrac{V_Y}{R_Y} \\ I_{C1} + I_{C2} = 2I_1 \end{cases}$$

Hence: $\tanh\left(\dfrac{V_D}{2V_T}\right)=\dfrac{V_Y}{R_Y I_1}$

For Q_3-Q_4 : $\begin{cases} I_{C3}-I_{C4}=2\dfrac{V_X}{R_X} \\ I_{C3}+I_{C4}=2I_1 \end{cases}$ and for Q_5-Q_6 :

$$\begin{cases} I_{C6}-I_{C5}=I_{C3}\tanh\left(\dfrac{V_D}{2V_T}\right) \\ I_{C6}+I_{C5}=I_{C3} \end{cases}$$

For Q_7-Q_8 : $\begin{cases} I_{C7}-I_{C8}=I_{C4}\tanh\left(\dfrac{V_D}{2V_T}\right) \\ I_{C7}+I_{C8}=I_{C4} \end{cases}$

Hence: $\begin{cases} I_{C6}=\dfrac{I_{C3}}{2}\left[1+\tanh\left(\dfrac{V_D}{2V_T}\right)\right] \\ I_{C5}=\dfrac{I_{C3}}{2}\left[1-\tanh\left(\dfrac{V_D}{2V_T}\right)\right] \end{cases}$ and $\begin{cases} I_{C7}=\dfrac{I_{C4}}{2}\left[1+\tanh\left(\dfrac{V_D}{2V_T}\right)\right] \\ I_{C8}=\dfrac{I_{C4}}{2}\left[1-\tanh\left(\dfrac{V_D}{2V_T}\right)\right] \end{cases}$

Since

$$I_A=I_{C6}+I_{C8}=\frac{I_{C3}+I_{C4}}{2}+\frac{I_{C3}-I_{C4}}{2}\tanh\left(\frac{V_D}{2V_T}\right)=I_1+\frac{V_X}{R_X}\tanh\left(\frac{V_D}{2V_T}\right)=I_1+\frac{V_X}{R_X}\frac{V_Y}{R_Y I_1}$$

$$I_B=I_{C5}+I_{C7}=\frac{I_{C3}+I_{C4}}{2}-\frac{I_{C3}-I_{C4}}{2}\tanh\left(\frac{V_D}{2V_T}\right)=I_1-\frac{V_X}{R_X}\tanh\left(\frac{V_D}{2V_T}\right)=I_1-\frac{V_X}{R_X}\frac{V_Y}{R_Y I_1}$$

So, finally: $I_A-I_B=2\dfrac{V_X}{R_X}\dfrac{V_Y}{R_Y I_1}$

Application to signal product: decompose difference $I_A - I_B$ when $V_X = V_1 \cos(2\pi f_X t)$; $V_Y = V_1 \cos(2\pi f_Y t)$:

$$I_A - I_B = 2\frac{V_1^2}{R_X R_Y I_1}\cos(2\pi f_X t)\cos(2\pi f_Y t)$$
$$= \frac{V_1^2}{R_X R_Y I_1}\left[\cos(2\pi(f_X + f_Y)t) + \cos(2\pi(f_X - f_Y)t)\right]$$

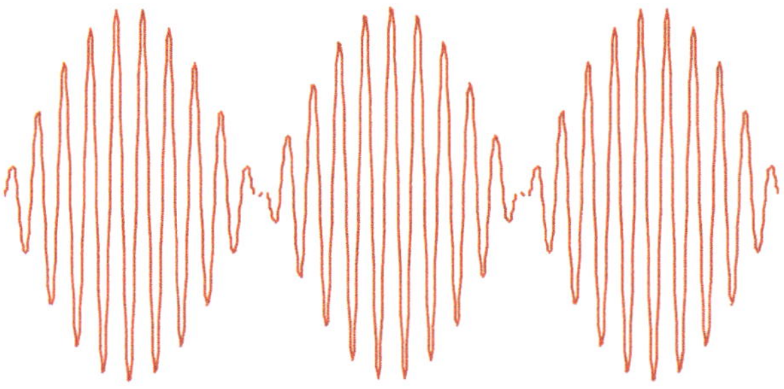

Case $f_x/f_y = 20$

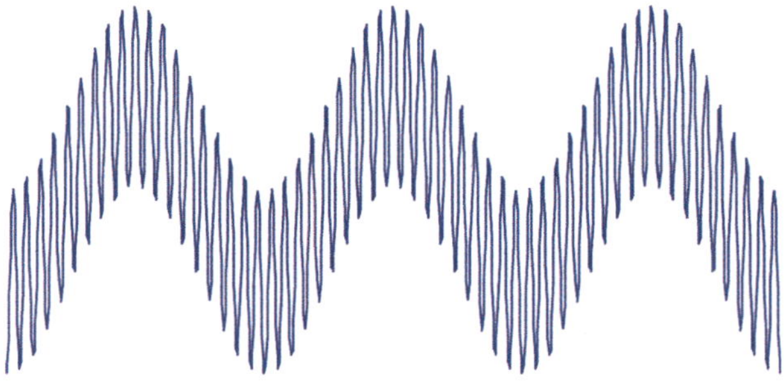

Case $f_x/f_y = 10/9$

2.9.3. *Amplifiers with rest current zero or low relative to nominal current*

i) Using letter notation for electrical quantities, determine the static (or rest, that is at $e_g = 0$) emitter currents I_{E1} and I_{E2} in the AB class circuit below, assuming zero to be the dynamic resistance of diodes D1, D2 and BE and all thresholds at U_0, with transistor current gain at β.

ii) Draw the output signal when $e_g = 10\text{V} \sin(2\pi f_0 t)$. What is the condition on E if the signal at the R_C terminals is not limited? What happens if we remove the diodes and resistances R_A and R_E short-circuits?

iii) In the following circuitry, load Z_C is made up of resistance R, an inductance L and a capacitance C in parallel, giving an overvoltage coefficient Q_0.

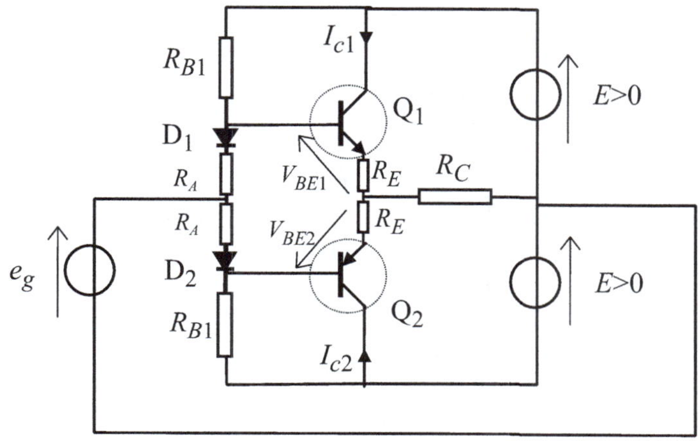

The generator supplies rectangular impulses such that $i_C(t)$ is also of rectangular shape, duty cycle ratio a, of frequency $\dfrac{1}{2\pi\sqrt{LC}}$ and maximum value I_{Cm}. Using the Fourier series decomposition of $i_C(t)$ in real terms, write $v_{CE}(t)$. In what conditions can we obtain the maximum amplitude for the fundamental frequency? And the maximum yield for the fundamental frequency?

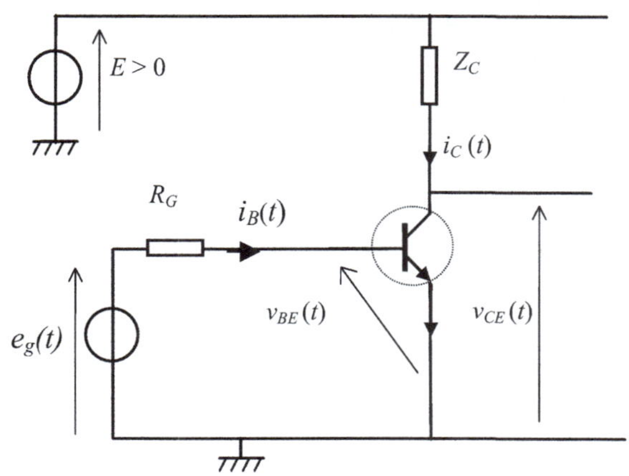

Answer:

i) Accounting for types npn of Q1 and pnp of Q2, and of $e_g = 0$, we obtain the equivalent circuit below:

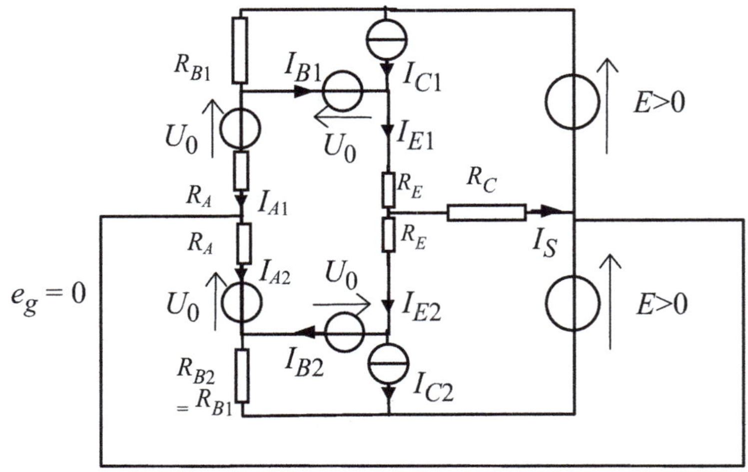

We write the loop analysis accounting for $I_B = I_E/(\beta + 1)$ and for $R_{B1} = R_{B2} = R_B$:

$$\begin{cases} R_B\left[I_{A1} + \dfrac{I_{E1}}{\beta+1}\right] + R_A I_{A1} + U_0 = E & (1) \\ R_B\left[I_{A2} + \dfrac{I_{E2}}{\beta+1}\right] + R_A I_{A2} + U_0 = E & (2) \\ R_A I_{A1} - R_E I_{E1} - R_C I_S = 0 & (3) \\ R_A I_{A2} - R_E I_{E2} + R_C I_S = 0 & (4) \end{cases}$$

and the node equation: $I_{E1} - I_{E2} - I_S = 0$ or alternatively $I_S = I_{E1} - I_{E2}$.

Writing $R_D = R_E/(\beta + 1)$ and $I_{AD} = I_{A1} - I_{A2}$, this system is reduced to a homogenous system of two equations with two unknowns by performing the subtractions (1) – (2) and (3) – (4):

$$\begin{bmatrix} R_B + R_A & R_D \\ R_A & -(R_E + 2R_C) \end{bmatrix} \begin{bmatrix} I_{AD} \\ I_S \end{bmatrix} = \begin{bmatrix} 0 \\ 0 \end{bmatrix}$$

Since both terms of the determinant do not have the same sign, it is necessarily non-zero and hence we must have $I_{AD} = 0$ and $I_S = 0$, so $I_{A1} = I_{A2}$ and $I_{E1} = I_{E2}$.

Using (2) and (3), we obtain the rest current:

$$I_E = \frac{E - U_0}{R_B \frac{R_E}{R_A} + \frac{R_B}{\beta + 1} + R_E}$$

This current is not zero and so still maintains both transistors in light conduction (AB class).

ii) If R_E is very low relative to R_C, we will have $R_E\, I_E << R_C\, I_S$ and $R_A\, I_A << R_C\, I_S$ as soon as $e_g \neq 0$. So $e_g = 10\text{ V} \sin(2\pi f_0 t) \approx R_C I_S$ (common collector assembly). In order for the signal at the R_C terminals not to be limited, the transistors must never be saturated, so $E > e_{gmax} = 10$ V.

If we remove the diodes and resistances R_A and R_E, the transistors will remain cutoff as long as $| e_g | < U_0$, which would lead to a distortion on passage of voltage U_S at the R_C terminals approaching zero, giving $U_S - U_0$ for $e_g > U_0$ and $U_S + U_0$ for $e_g < -U_0$.

iii) C class circuit: Fourier series decomposition of $i_C(t)$ to real terms gives us:

$$i_C(t) = I_{Cm}\left[a + \sum_{n=1}^{\infty} \frac{2\sin(n\pi\, a)}{n\pi} \cos(2\pi\, n\, f_1 t)\right]$$

where $f_1 = \frac{1}{T} = \frac{1}{2\pi\sqrt{LC}}$

in order to excite the RLC parallel circuit to its resonance frequency using the current's fundamental component.

Accordingly, for each harmonic frequency nf_1, there are voltages whose alternative component $v_{CEn}(t)$ is determined by the complex impedance of the circuit $Z(nf_1)$ with argument φ_n and the current component of the same frequency, with $\phi_n = -\text{Arctg}\left[Q_0\left(n - \frac{1}{n}\right)\right]$ and $|Z(nf_1)| = \frac{R}{\sqrt{1+\left[Q_0\left(n-\frac{1}{n}\right)\right]^2}}$, which is R for $n = 1$ and then decreases:

$$v_{CEn}(t) = -|Z(nf_1)|\, I_{Cm}\, \frac{2\sin(n\pi\, a)}{n\pi} \cos(2\pi\, n\, f_1 t + \varphi_n)$$

Since the inductor connects the collector to the power supply source which provides the DC voltage E, the average collector-emitter voltage is also E. Hence, the whole collector-emitter voltage becomes:

$$V_{CE_n}(t) = E - \sum_{n=1}^{\infty} v_{CE_n}(t) = E - \sum_{n=1}^{\infty} |Z(nf_1)|\, I_{C_m}\, \frac{2\sin(n\pi\, a)}{n\pi} \cos(2\pi\, n\, f_1 t + \phi_n)$$

which means that $V_{CE}(t)$ can reach near zero on transistor saturation, a peak amplitude E and a peak to peak variation $2E$. For fundamental order $n = 1$, this situation corresponds to:

$$E = RI_{Cm}\, \frac{2\sin(\pi\, a)}{\pi}$$

Power in load R is then $\frac{1}{R}\left(\frac{E}{\sqrt{2}}\right)^2 = \frac{E^2}{2R}$, while the power delivered by the supply is aEI_{Cm}. We can deduce the yield $\gamma = \frac{E}{2RaI_{Cm}} = \frac{\sin(\pi a)}{\pi a}$ which tends toward 1, that is 100%, when the duty cycle ratio a tends toward 0.

This is not completely true due to harmonics that dissipate some power in resistance. We can compare the power dissipated by each n harmonic relative to the power of the fundamental order $n = 1$ by working out the ratio of the square of voltage amplitudes, which themselves are equal to the product of the impedance modulus by the current amplitude at each frequency:

$$\left(\frac{|Z(nf_1)|\sin(n\pi a)}{|Z(f_1)|\sin(\pi a)}\right)^2 = \frac{1}{1+Q_0^2\left[n-\frac{1}{n}\right]^2}\left[\frac{\sin(n\pi a)}{\sin(\pi a)}\right]^2$$

ratio that does not exceed 2% when $Q_0 \geq 5$ and $n = 2$ or larger.

Accordingly, this type of operation is prized in radiofrequency where yield has to be maximized.

2.9.4. *Cascode circuit and frequency responses of transistor amplifiers*

Numerical values identical for Q_1 and Q_2, typical for transistor BC237 operating at $I_{C1} = I_{C2} = 5$ mA with $R_C = 500\ \Omega$; $\beta = 200$; $r_{BB'} = 140\ \Omega$; $\eta V_T = 30$ mV; $\tau_{B'E} = r_{B'E}\, C_{B'E} = 160$ ns; $C_{B'C} = 3$ pF, in the following circuit.

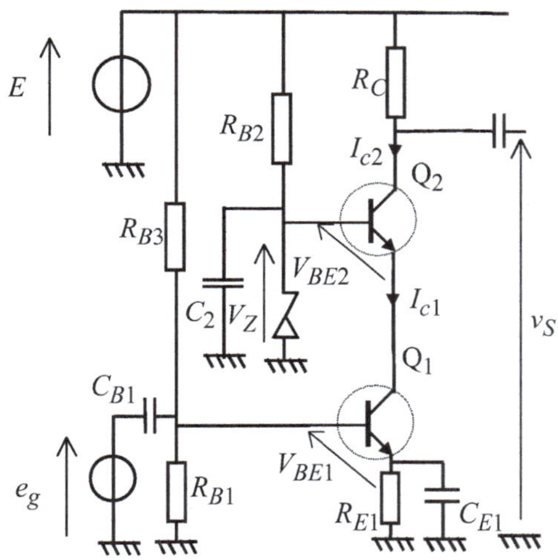

The impedance of the capacitors present inside the assembly is considered negligible. We also disregard the conductance of R_{B3} // R_{B1} vis-à-vis the reverse of the Q_1 input dynamic resistance.

Find the equivalent (linearized) small signal circuit of stage Q_1, accounting for the transistor capacitances and provisionally replacing stage Q_2 by a single R_C resistance. Let us assume $r_{CE} >> r_{B'E}$ and $r_{CE} >> R_C$ (case of common emitter circuit in which we disregard conductance $1/\, r_{CE}$).

Determine the expression of voltage gain $v_S \,/\, e_g$ according to circular frequency ω (or requency f) and its cutoff frequency f_{c1} by performing the Miller effect approximation.

Same question for the complete cascode circuit (f_{c2}). What is the circuit type for stage Q_2?

Answer:

i) In the equivalent dynamic circuit, we have $v_{B'E1} = r_{B'E}\, i_{B1}$:

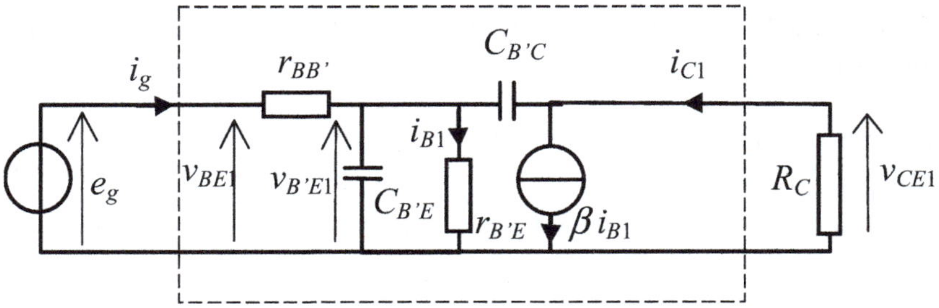

By disregarding the current in r_{CE} relative to i_{C1}:

$$v_{CE1} = -R_C i_{C1} = -R_C\left[\beta i_{B1} + jC_{B'C}\omega(v_{CE1} - v_{B'E1})\right]$$

and $e_g = r_{BB'} i_g + r_{B'E} i_{B1}$.

Since the node equation at base B', when accounting for the current division ratio in $r_{B'E}$ and $C_{B'E}$, gives:

$$i_g = \left[1 + j\, r_{B'E}\, C_{B'E}\omega\right] i_{B1} - j\, C_{B'C}\omega(v_{CE1} - v_{B'E1})$$

we write $A_v = -\left.\dfrac{v_{CE1}}{v_{B'E1}}\right|_{\omega=0} = \dfrac{\beta R_C}{r_{B'E}}$ disregarding the current in $C_{B'C}$ relative to i_{C1}. We then have:

$$\begin{cases} v_{CE1} = -R_C\left[\beta - j(1+A_v)C_{B'C} r_{B'E}\omega)\right] i_{B1} \\ e_g = r_{BB'} i_{B1}\left[1 + j r_{B'E} C_{B'E}\omega + j(1+A_v)C_{B'C} r_{B'E}\omega)\right] + r_{B'E} i_{B1} \end{cases}.$$

We obtain (with $A_v >> 1$ at the numerator):

$$\frac{v_{CE1}}{e_g} = -A_v \frac{r_{B'E}}{r_{B'E} + r_{BB'}} \frac{1 - jR_C C_{B'C}\omega}{1 + j\dfrac{r_{B'E} r_{BB'}}{r_{B'E} + r_{BB'}}\omega\left[C_{B'E} + (1+A_v)C_{B'C}\right]}$$

The effect of the time constant located at the numerator can be disregarded since it is much lower than that of the denominator, which will determine the cutoff frequency of the voltage gain. Numerical verification:

$$r_{B'E} \approx \frac{V_T}{I_B} = \frac{\beta V_T}{I_C} = 1200\ \Omega \quad \text{and} \quad r_{B'E} // r_{BB'} = 125\ \Omega;$$

$$A_v = \frac{\beta R_C}{r_{B'E}} = 83; \; C_{B'E} = \frac{\tau_{B'E}}{r_{B'E}} = 133\ \text{pF}$$

Hence, $C_{B'E} + (1 + A_v)C_{B'C} = 385\,\text{pF}$ so 100 times greater than $C_{B'C}$ due to the effect of gain on the equivalent capacitance taken on input (Miller effect). The time constant of the numerator is consequently 1.5 ns, negligible relative to that of the denominator that is 48 ns, which essentially amounts to neglecting the capacitive current in $C_{B'C}$ relative to $\beta\, i_{B1}$. Accordingly, we will remove the imaginary term from the numerator in what follows.

However, the voltage at the R_C terminals is actually determined by the flow of $\beta\, i_{B1}$ into R_C in parallel with $C_{B'C}[1+1/A_v] \approx C_{B'C}$, which leads to $A_v = \frac{\beta R_C}{r_{B'E}} \frac{1}{1 + j\omega C_{B'C} R_C}$. The consistent approximation of taking A_v at $\omega = 0$

is only valid if $C_{B'C} R_C << \frac{r_{B'E} r_{BB'}}{r_{B'E} + r_{BB'}} \left[C_{B'E} + (1 + A_v)C_{B'C} \right]$, which is the case ($R_C\, C_{B'C}$ = 1.5 ns), and if we only want to determine the first cutoff frequency. We write:

$$f_{c1} = \frac{\omega_{c1}}{2\pi} = \frac{1}{2\pi \frac{r_{B'E} r_{BB'}}{r_{B'E} + r_{BB'}} \left[C_{B'E} + (1 + A_v)C_{B'C} \right]} = 3.3\ \text{MHz}$$

ii) The circuit for stage Q_2 is of the common base type since the base is grounded from a dynamic point of view with input occurring at the emitter. So, input impedance will be reduced and therefore decrease the Miller effect while the voltage gain is carried forward to the second stage.

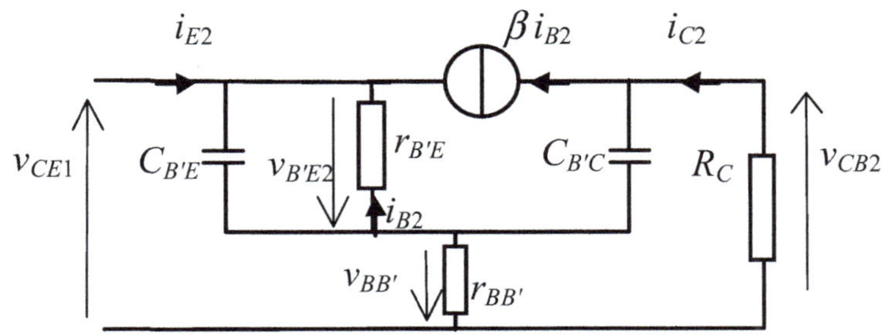

We have $v_{B'E2} = r_{B'E}\ i_{B2}$, $v_{CB2} = -\beta R_C i_{B2} = -\dfrac{\beta R_C}{r_{B'E}} v_{B'E2}$, as before disregarding the current in $C_{B'C}$ and at node B':

$$-\frac{v_{B'E2}}{r_{B'E}} - jC_{B'E}\omega v_{B'E2} + \frac{v_{BB'}}{r_{BB'}} + jC_{B'C}\omega\left[-\frac{\beta R_C}{r_{B'E}} v_{B'E2} + v_{BB'}\right] = 0$$

Hence $v_{CE1} = -\left(v_{B'E2} + v_{BB'}\right) = -r_{B'E} i_{B2}\left[1 + \dfrac{r_{BB'}}{r_{B'E}} \dfrac{1 + j\omega r_{B'E}\left(C_{B'E} + A_v C_{B'C}\right)}{1 + j\omega\, r_{BB'} C_{B'C}}\right]$

Current i_{E2} is $i_{E2} = -(\beta+1)i_{B2} - j\omega C_{B'E} v_{B'E2} = -\left[(\beta+1) + j\omega C_{B'E} r_{B'E}\right] i_{B2}$

Hence, the load impedance of the first stage is:

$$Z_c = \frac{v_{CE1}}{i_{E2}} = \frac{r_{B'E} + r_{BB'} \dfrac{1 + j\omega r_{B'E}\left(C_{B'E} + A_v C_{B'C}\right)}{1 + j\omega C_{B'C} r_{BB'}}}{\beta + 1 + j\omega C_{B'E} r_{B'E}}$$

which again gives $\dfrac{r_{B'E} + r_{BB'}}{\beta + 1}$ at low frequency and that can be approximated by:

$$Z_c = \frac{r_{B'E} + r_{BB'}}{\beta+1} \frac{\left[1 + j\omega \dfrac{r_{B'E} r_{BB'}}{r_{B'E} + r_{BB'}}\left(C_{B'E} + A_v C_{B'C}\right)\right]}{1 + j\omega \dfrac{C_{B'E} r_{B'E}}{\beta + 1}}$$

$$\approx \frac{r_{B'E} + r_{BB'}}{\beta+1}\left[1 + j\omega \frac{r_{B'E} r_{BB'}}{r_{B'E} + r_{BB'}}\left(C_{B'E} + A_v C_{B'C}\right)\right]$$

as the time constant of denominator $\dfrac{C_{B'E}r_{B'E}}{\beta+1} = 0.8$ ns can be neglected.

The second level gain is given, disregarding the time constant $r_{BB'}\,C_{B'C}$ = 0.42 ns, by:

$$\frac{v_{CB2}}{v_{CE1}} = \frac{-\beta R_C i_{B2}}{-r_{B'E}i_{B2}\left[1+\dfrac{r_{BB'}}{r_{B'E}}\dfrac{1+j\omega r_{B'E}\left(C_{B'E}+A_v C_{B'C}\right)}{1+j\omega\, r_{BB'}C_{B'C}}\right]}$$

$$= \frac{r_{B'E}}{r_{B'E}+r_{BB'}}\frac{A_v}{1+j\omega\dfrac{r_{B'E}r_{BB'}}{r_{B'E}+r_{BB'}}\left(C_{B'E}+A_v C_{B'C}\right)}$$

scarcely differing in absolute value from the gain of the first stage loaded by R_C.

To obtain the total gain we must combine both gains, accounting for Z_C instead of R_C as a load for the first stage. Since we want the first cutoff frequency, we can replace Z_C by its real part $\dfrac{r_{B'E}+r_{BB'}}{\beta+1}$ in order to calculate the new gain denominator of the first stage A_v' (the whole expression of Z_C would provide a corrective term of the second degree to the denominator):

$$A_v' = -\left.\frac{v_{CE1}}{v_{B'E1}}\right|_{\omega=0,Z_c} = \frac{\beta\left(r_{B'E}+r_{BB'}\right)}{r_{B'E}\left(\beta+1\right)} \approx \frac{r_{B'E}+r_{BB'}}{r_{B'E}},$$

which then becomes just greater than 1 while it had been 83. So we can replace $(1+A_v')C_{B'C}$ by $2C_{B'C}$.

However, we must take the full expression of Z_C for the numerator of the new first stage gain since it contains a first-degree polynomial in ω:

$$A_{vtotal} = \frac{v_{CB2}}{v_{CE1}}\frac{v_{CE1}}{e_g} = -\frac{\left[1+j\omega\dfrac{r_{B'E}r_{BB'}}{r_{B'E}+r_{BB'}}\left(C_{B'E}+A_v C_{B'C}\right)\right]}{1+j\dfrac{r_{B'E}r_{BB'}}{r_{B'E}+r_{BB'}}\omega\left[C_{B'E}+2C_{B'C}\right]}$$

$$\frac{r_{B'E}}{r_{B'E}+r_{BB'}}\frac{A_v}{1+j\omega\dfrac{r_{B'E}r_{BB'}}{r_{B'E}+r_{BB'}}\left(C_{B'E}+A_v C_{B'C}\right)}$$

We then simplify and finally obtain:

$$A_{vtotal} = \frac{v_{CB2}}{v_{CE1}} \frac{v_{CE1}}{e_g} = -\frac{r_{B'E}}{r_{B'E} + r_{BB'}} \frac{A_v}{1 + j\dfrac{r_{B'E} r_{BB'}}{r_{B'E} + r_{BB'}} \omega \left[C_{B'E} + 2C_{B'C} \right]}$$

The new cutoff frequency is notably higher since the intervening capacitance is now $C_{B'E} + 2C_{B'C}$ instead of $C_{B'E} + (1 + A_v)C_{B'C}$, giving a time constant of 17.3 ns:

$$f_{c2} = \frac{\omega_{c2}}{2\pi} = \frac{1}{2\pi \dfrac{r_{B'E} r_{BB'}}{r_{B'E} + r_{BB'}} \left[C_{B'E} + 2C_{B'C} \right]} = 9.16 \text{ MHz}$$

The low frequency (LF) gain of the overall circuit remains exactly the same as that of the first stage if it had been loaded by R_C:

$$A_{vtotal} = \frac{v_{CB2}}{v_{CE1}} \frac{v_{CE1}}{e_g} = -A_v \frac{r_{B'E}}{r_{B'E} + r_{BB'}} = 74$$

3

Field Effect Transistors and Applications

3.1. Operating principle of junction field effect transistors (JFET and MESFET types)

The channel (n-type) represented below (Figure 3.1) by the non shaded zone has a width depending on abscissa x and voltages V_{GS} and V_{DS} and is traversed by drain current I_D, as shown in Figure 3.1.

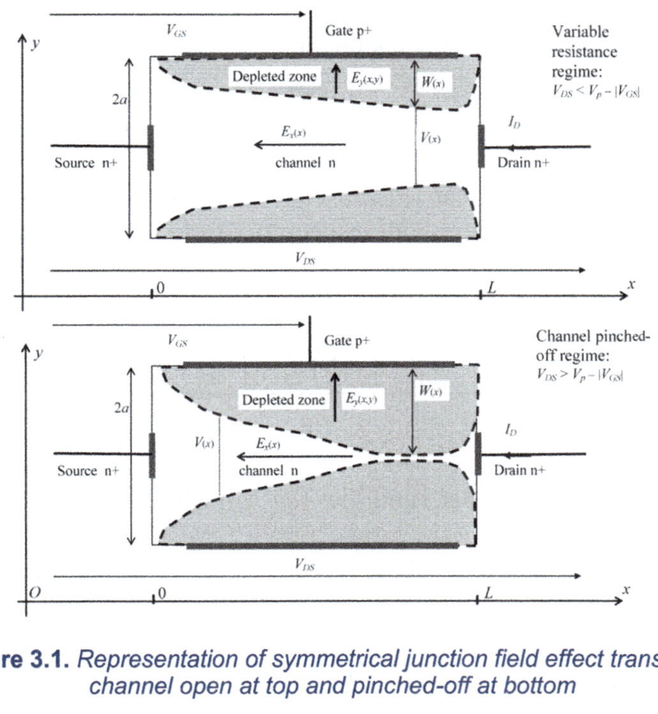

Figure 3.1. *Representation of symmetrical junction field effect transistor; channel open at top and pinched-off at bottom*

Here, the structure is taken as symmetrical in order not to account for the influence of the substrate; however, the results are, nonetheless, applicable for real, asymmetrical components built on a semi-isolating substrate. The width of the depletion zones is calculated for the pn diode (Chapter 1, section 1.4.3), while accounting for the difference in potential between the gate and the channel, that is $V_{GS} - V(x)$; with $V(x) = 0$ at the source and $V(x) = V_{DS} > 0$ at the drain. The gate is reverse biased by $V_{GS} < 0$ and $N_A >> N_D$, so that: $W(x) = \sqrt{\dfrac{2\varepsilon}{eN_D}}\sqrt{\Phi - V_{GS} + V(x)} = \sqrt{\dfrac{2\varepsilon}{eN_D}}\sqrt{\Phi + |V_{GS}| + V(x)}$, where Φ is the diffusion potential or built-in potential of the gate–channel junction (Chapter 1, section 1.4.1).

In the following, drain current will be calculated in half the height (dimension a) of the structure drawn in Figure 3.1. The electric field $E_x(x)$ that is predominant in the channel determines conduction current I_D in two different ways according to whether the channel is pinched-off or not:

1) In a variable resistance regime, occurring in an open channel, the channel's resistance is dependent on its fixed width Z and its height $h(x) = a - W(x)$ and we can calculate the conduction current by $I_D = eZhN_D\,\mu_n \dfrac{dV}{dx}$. If we differentiate equation $h(x) = a - \sqrt{\dfrac{2\varepsilon}{eN_D}}\sqrt{\Phi - V_{GS} + V(x)}$, we can obtain dV according to dh that is carried subsequently into the expression of $I_D\,dx$. By integrating both members with respect to x from 0 to L and with respect to h from $h(0)$ to $h(L)$, respectively, in cases where $h(L)$ remains non-zero (non-pinched-off channel), we have:

$$I_D = I_P\left[3\frac{V_{DS}}{V_P} - 2\frac{\left(V_{DS} + |V_{GS}| + \Phi\right)^{3/2} - \left(|V_{GS}| + \Phi\right)^{3/2}}{V_P^{3/2}}\right]$$

where $I_P = (e^2\mu_n Za^3N_D^2)/(6\varepsilon L)$ and $V_P = (ea^2N_D)/(2\varepsilon)$, the pinch-off voltage.

Channel conductance g_D is obtained by deriving I_D with respect to V_{DS}:

$$g_D = \frac{dI_D}{dV_{DS}} = 3\frac{I_P}{V_P}\left[1 - \frac{\left(V_{DS} + |V_{GS}| + \Phi\right)^{1/2}}{V_P^{1/2}}\right]$$

and the transistor can operate around $V_{DS} = 0$ in variable resistance regime controlled by V_{GS}.

2) At a given value of V_{GS}, maximum current is reached when V_{DS} cancels the derivative of I_D, that is g_D, for $V_{DSP} = V_P - \Phi - |V_{GS}|$. Current is then equal to: $I_{Dsat}(V_{GS}) = I_P\left[1-3\frac{|V_{GS}|+\Phi}{V_P}+2\left(\frac{|V_{GS}|+\Phi}{V_P}\right)^{3/2}\right]$ and this regime corresponds to canal pinching-off at point $x = L$.

For values of $V_{DS} > V_P - \Phi - |V_{GS}|$, the conduction model above is no longer valid in the pinched-off part of the channel, since $E_x(x)$ becomes very high and mobility μ_n decreases so that the speed of carriers $\mu_n E_x$ reaches a limit set by the semiconductor. Current then remains approximately constant and equal to $I_{Dsat}(V_{GS})$, the so-called saturation current.

This current is maximum for $V_{GS} = 0$ and is $I_{DSS} = I_P\left[1-3\frac{\Phi}{V_P}+2\left(\frac{\Phi}{V_P}\right)^{3/2}\right] \approx I_P$ since $\Phi \ll V_P$. For $|V_{GSoff}| = V_P - \Phi$, current I_{Dsat} becomes zero and the transistor is in an off-state. Accordingly, pinch-off limit V_{DSP} can also be expressed as $V_{DSP} = |V_{GSoff}| - |V_{GS}|$.

For $V_{DS} > V_{DSP}$, *the transistor is a current source controlled by* V_{GS} and we can calculate its transconductance g_m by deriving I_{Dsat} with respect to V_{GS}:

$$g_m = \frac{3I_P}{V_P}\left[1-\left(\frac{|V_{GS}|+\Phi}{V_P}\right)^{1/2}\right] \text{ where } g_m < g_{m\max} = \frac{3I_P}{V_P}$$

These transistors are, therefore, still conductive at $V_{GS} = 0$ and require voltage V_{GSoff} of opposite sign (here negative) to the V_{DS} value needed to reach pinched-off regime (or current saturation) in order to be off. Consequently, they are known as “depletion FETs” since a control voltage must be used in order for the majority carriers to be expelled from the channel, which is then depleted of carriers and ultimately insulating when $V_{GS} < V_{GSoff}$ as we can see in graph $I_{DSat}(V_{GS})$ of Figure 3.2. Metal-semiconductor FETs (MESFETs) operate in the same way but are built with a Schottky diode instead of the pn diode for the gate-channel junction.

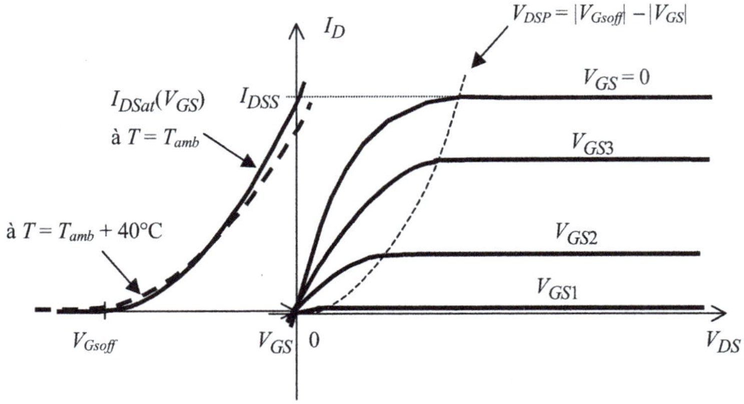

Figure 3.2. *Characteristic $I_{DSat}(V_{GS})$ and $I_D(V_{DS})$ of JFET or N-MESFET*

Characteristic $I_{DSat}(V_{GS})$ is frequently replaced by a parabolic approximation:

$$I_{DSat}(V_{GS}) = I_{DSS}\left[1 - \frac{V_{GS}}{V_{GSoff}}\right]^2$$

Furthermore, when V_{DS} becomes higher than $V_{DSP} = |V_{GSoff}| - |V_{GS}|$, the drain current typically continues to increase slightly and a parallel conductance g_{DS} on the drain current source is required to report this effect.

Characteristic $I_{DSat}(V_{GS})$ presents a point independent of temperature since I_{DSS} decreases as temperature increases due to reduced mobility, but $|V_{GSoff}| = V_P - \Phi$ increases as Φ decreases.

Gate currents I_G are limited to the reverse diode leakage currents and double every 10°C for silicon. However, they can increase substantially under high voltage V_{DS} (or V_{DG}) due to the ionization phenomenon at the border of the depletion zone (see Figure 3.3).

We can find p-channel junction field effect transistors (JFET) (all voltage and current signs inverted) mostly on silicon transistor and metal semiconductor field effect transistor (MESFET) on GaAs and InP, which include a Schottky gate-channel diode instead of a pn diode, only on n-type channels since electron mobility is much higher than for the holes in III–V semiconductors.

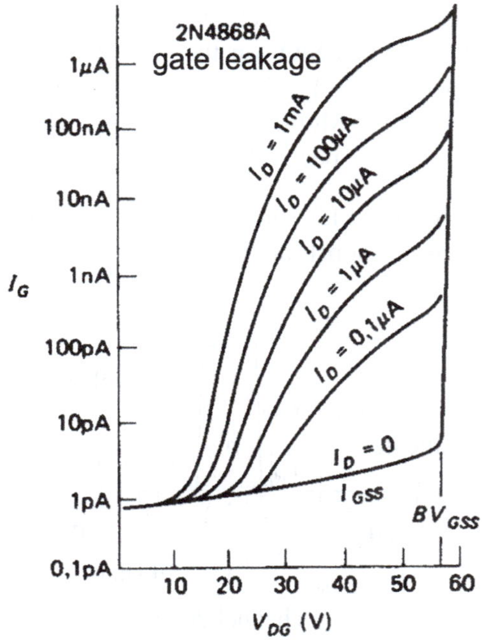

Figure 3.3. *Gate leakage current of a JFET with n-type silicion channel*

The symbols that represent depletion JFET (the solid line between drain and source appears as a conductive channel at $V_{GS} = 0$) are as follows (Figure 3.4).

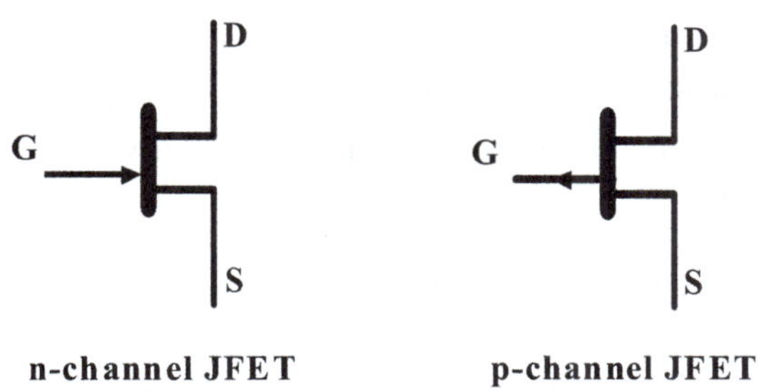

Figure 3.4. *Symbols for depletion JFETs*

3.2. Metal oxide semiconductor field effect transistors

In metal oxide semiconductor field effect transistors (MOSFET), the gate is made up of a metallically conductive material and separated from the channel by an insulator (silica SiO_2 on the silicon, or increasingly, another oxide with higher permittivity). These transistors are abbreviated to N-MOSFET (or NMOS) and P-MOSFET (or PMOS) when conduction in the channel is due to electrons, respectively holes.

For enhancement N-MOSFET (or NMOS), two n+ wells are implanted in a weakly doped (N_A) p-type substrate that comprises the channel.

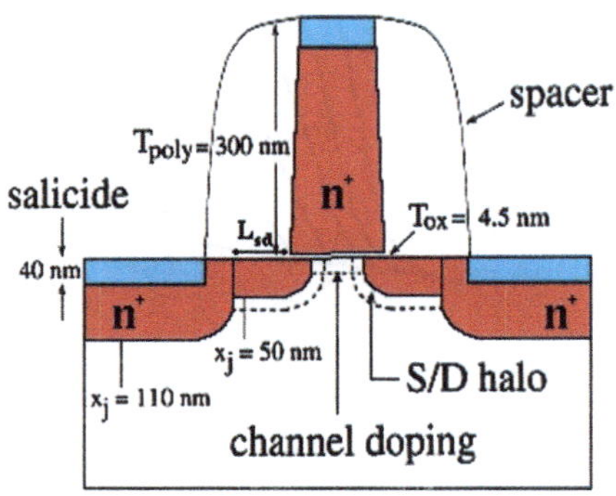

Figure 3.5. *Modern MOSFET (metallic electrodes at the top of n+ zones, gate at the center, source and drain at the borders)*

Consequently, from the drain to the source, we see two n+/p diodes end to end. The transistor is in an off-state in the absence of voltage applied to the gate or when $V_{GS} = 0$. This accounts for the name "enhancement transistor", since it only becomes conductive if mobile charges of the same type as those present in the drain and source wells are induced by capacitive effect into the channel, which corresponds to the inversion of the type of carriers in the channel.

To make the N-MOSFET conductive, we must apply a positive V_{GS} voltage in order to induce a total negative charge Q_s per unit surface in the channel and to establish a n-type route between drain and source by means of inversion: $Q_s(x) = -C_{ox}[V_{GS} - (V(x) + \Phi')]$, where C_{ox} is the capacitance of the insulator per unit of surface, Φ' is the difference of internal potential in the insulator-semiconductor structure and $V(x)$ is the potential in the channel, analogous to reverse voltage if the gate should be replaced by a diode.

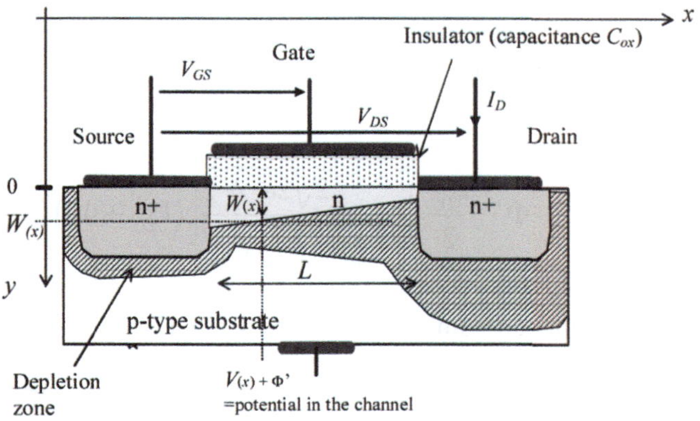

Figure 3.6. *n-Channel MOSFET in conductive regime because of inverted carriers (here electrons) induced in the channel of width W(x) (represented unshaded), non-pinched-off, for V_{GS} greater than the threshold voltage V_T. The shaded area features the depletion zone*

In order to obtain the charge of mobile electrons per unit of surface $Q_n(x)$, we must subtract the negative charge $-\sqrt{2e\varepsilon N_A[V(x)+\Phi']}$ due to the acceptors in concentration N_A on the channel depth $W(x)$, since these are fixed charges that do not participate in conduction. Hence, we arrive at:

$$Q_n(x) = -C_{ox}[V_{GS} - (V(x) + \Phi')] + \sqrt{2e\varepsilon N_A[V(x)+\Phi']}$$

For a channel of width Z and length L, conductance is provided by the parallel connection (or the summation) of all the conductance from $y = 0$ to $W(x)$:

$$G = \frac{Z}{L}\int_0^{W(x)} e\mu_n n(y)dy = \frac{\mu_n Z}{L}|Q_n(x)|$$

Hence, we deduce the voltage drop $dV = I_D\, dR$ in an element of channel length dx:

$$dV = I_D dR = I_D \frac{1}{G}\frac{dx}{L} = \frac{I_D dx}{\mu_n Z |Q_n(x)|}$$

By substituting the expression of $|Q_n(x)|$ into the previous equation and integrating $I_D\,dx$ from the source ($x = 0$, $V(0) = 0$) to the drain ($x = L$, $V(L) = V_{DS}$), we obtain I_D:

$$I_D = \frac{Z}{L}\mu_n C_{ox}\left\{\left(V_{GS} - \Phi' - \frac{V_{DS}}{2}\right)V_{DS} - \frac{2\sqrt{2e\varepsilon N_A}}{3C_{ox}}\left[\left(V_{DS} + \Phi'\right)^{3/2} - \Phi'^{3/2}\right]\right\}$$

Conductance g_D of the channel is obtained by deriving I_D with respect to V_{DS}: $g_D = \frac{dI_D}{dV_{DS}} = \frac{Z}{L}\mu_n C_{ox}\left\{V_{GS} - \Phi' - V_{DS} - \frac{\sqrt{2e\varepsilon N_A}}{C_{ox}}\sqrt{V_{DS} + \Phi'}\right\}$

The transistor can thus operate around $V_{DS} = 0$ as a variable resistance controlled by V_{GS} with a conductance varying linearly with V_{GS}. The transistor becomes an open circuit when the conductance of the channel at $V_{DS} = 0$ becomes zero, that is when V_{GS} is equal to the (positive) threshold voltage: $V_T = \Phi' + \frac{\sqrt{2e\varepsilon N_A}}{C_{ox}}\sqrt{\Phi'}$.

2) When V_{DS} increases, current I_D reaches its maximum (when its derivative is cancelled, that is when $g_D = 0$) and becomes approximately constant since the conduction model above no longer applies and it is the speed of carriers $\mu_n\ E_x$ that becomes constant. This current I_{Dsat} corresponds to the case where $g_D = 0$ for $V_{DS} = V_{DSP}$ (channel pinched-off at $x = L$), that is for $V_{GS} = \Phi' + V_{DSP} + \frac{\sqrt{2e\varepsilon N_A}}{C_{ox}}\sqrt{V_{DSP} + \Phi'}$, which defines:

$$V_{DSP} = V_{GS} - \Phi' - \frac{e\varepsilon N_A}{C_{ox}^2}\left[\sqrt{1 + \frac{2V_{GS}C_{ox}^2}{e\varepsilon N_A}} - 1\right]$$

(which is cancelled for $V_{GS} = V_T$).

In channel pinched-off regime, the transistor is thus a current source controlled by V_{GS}.

Carrying V_{DSP} into I_D provides us with the expression of $I_{Dsat}(V_{GS})$. However, in most cases, voltage $\frac{e\varepsilon N_A}{C_{ox}^2}$ is less than 1 mV and much less than V_{GS}, which makes the last term of V_{DSP} negligible. We now have $V_T = \Phi'$ and $V_{DSP} = V_{GS} - \Phi'$; hence, the approximately parabolic and widely used relation is given as (including mobility measurement purposes):

$$I_{DSat}(V_{GS}) = \frac{Z}{2L}\mu_n C_{ox}(V_{GS} - V_T)^2$$

Transconductance g_m in this regime is then obtained by deriving $I_{Dsat}(V_{GS})$ with respect to V_{GS}, which we write simply as:

$$g_m = \frac{Z}{L}\mu_n C_{ox}(V_{GS} - V_T)$$

In order to account for the slight increase in drain current when V_{DS} becomes greater than V_{DSP}, a conductance g_{DS} has to be positioned in parallel on the current source.

Current I_D is not strictly zero for $V_{GS} < V_T$ due to the diffusion current that was neglected in the calculation, and which actually decreases exponentially with $V_{GS} - V_T$.

Like in JFETs, characteristic $I_{DSat}(V_{GS})$ passes through a point independent of temperature.

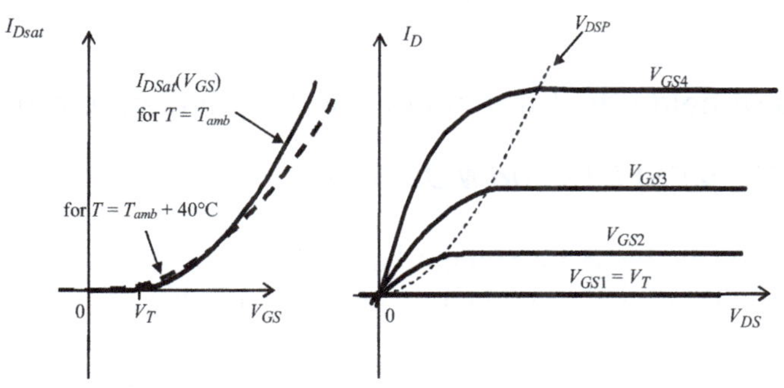

Figure 3.7. *n-Channel MOSFET characteristic $I_{DSat}(V_{GS})$ and $I_D(V_{DS})$*

When the MOSFET is an open circuit, we must account for the substrate potential if it is not connected to the source. Indeed, in such cases, the gate–channel interface potential is no longer dependent on the drain and source potentials but rather on gate potential V_G and substrate potential V_C since the channel is insulated from the source and the drain. In practice, we can consider that the channel is at a potential approaching V_C and can replace V_{GS} by V_{GC} in order to verify the off-state that is achieved if $V_{GC} < V_T$ for a NMOS or $V_{GC} > V_T$ for a P-MOSFET (or PMOS).

The gate current is very low (typically from 0.1 to 100 pA) and has little relationship to the applied voltages but does increase with temperature.

To obtain a depletion N-MOSFET, the channel is slightly doped with n-type impurities close to the interface with the insulator, which makes it conductive at $V_{GS} = 0$. We must then apply a voltage $V_{GS} < 0$ to make the transistor an open circuit by depleting the channel, with the threshold voltage V_T itself becoming negative.

In order to characterize the capability of a MOSFET to efficiently control the drain current variations, we generally use the transconductance per unit of channel length, that is g_m/Z (often in S/mm), which is proportional to μ_n/L in accordance with the last expression. Since ZC_{ox} is the capacitance per unit of channel length (as C_{ox} is the capacitance per unit surface), a relevant figure of merit for evaluating the cut-off frequency is composed of the inverse of the time constant $g_m/(Z^2C_{ox})$, proportional to μ_n/ZL. This explains the desire to reduce dimensions, in particular the channel length L, which all semiconductor manufacturers are attempting to assist in the production of microprocessors, telecommunications circuits, etc., and to increase mobility as well (see below).

3.3. Types of field effect transistors and equivalent circuits

3.3.1. *JFETs, MESFETs and MOSFETs*

The signs of the blocking voltage V_{GSoff} for JFETs and threshold voltage V_T for MOSFETs are indicated below together with that of voltages V_{DSP} in pinched-off (or saturation) regime in which a current source between drain and source is the most suitable electrical model. Nonetheless, field effect transistors are symmetrical in construction and are traversed by a conduction current that uses only one type of carrier into the channel in conductive state. The drain current I_D can thus flow either in the positive or negative direction according to the sign of V_{DS}. Such a property can give rise to significant applications under the variable resistance regime for which the channel is not pinched-off and consequently not modeled by a current source but by a simple resistance (or conductance) dependent on V_{GS}.

In the following figures, the symbols are associated with voltage signs V_{DS} in pinched-off conditions but they can be reversed under variable resistance regime. The channel is symbolized by a thick line, which is continuous when the transistor is conductive at $V_{GS} = 0$ (depletion transistor) or interrupted when the transistor is off at $V_{GS} = 0$ (enhancement transistor), as shown in Figure 3.8.

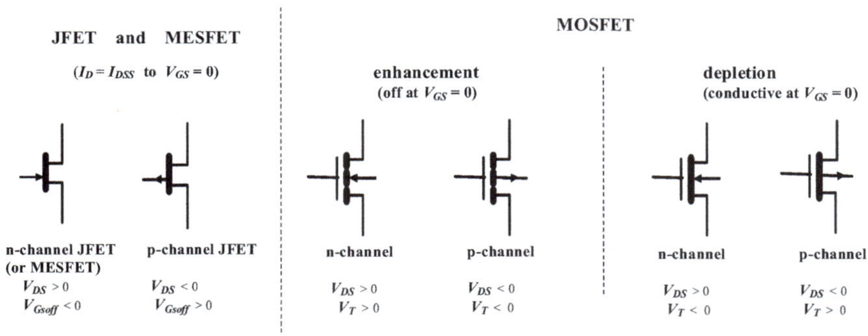

Figure 3.8. *Symbols of the various field effect transistors*

The equivalent linear circuit for small signals depends on the operating conditions (pinched-off channel or not) and is sketched in Figure 3.9. In most cases, we can disregard the gate current and therefore the gate conductance too.

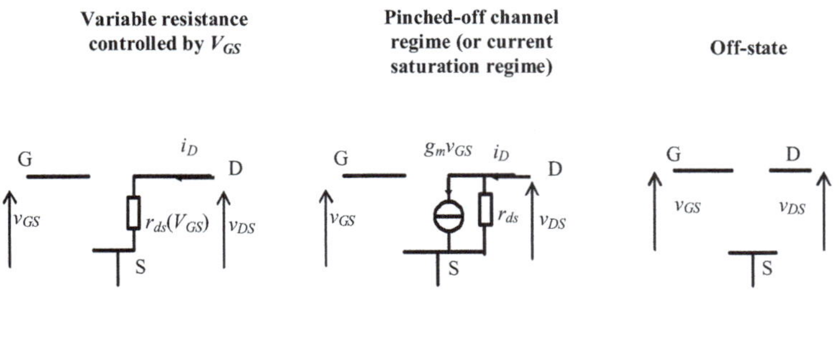

Figure 3.9. *Equivalent electrical circuits in linearized static or low-frequency dynamic regimes*

Under high-frequency or under rapid switching regime, it becomes necessary to account for junction or gate capacitances. An approximate model consists of replacing the gate-channel capacitances, spread along the entire channel length by two capacitances, gate-source and gate-drain. An output capacitance, located between source and drain, is also typically present due to the connections and the influence of the substrate.

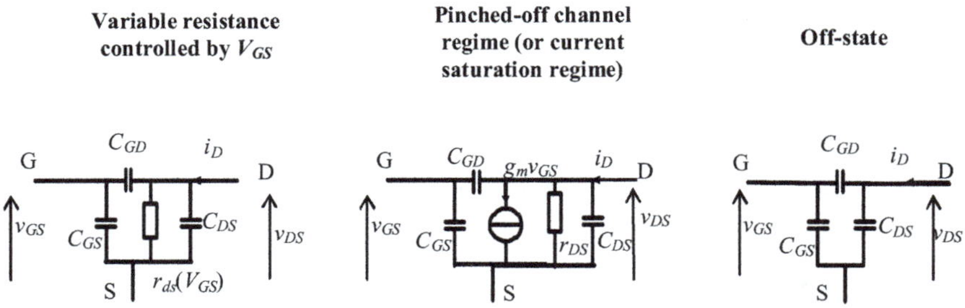

Figure 3.10. *Equivalent electrical circuits under dynamic high-frequency or rapid switching regimes*

3.3.2. *Other field effect transistors*

3.3.2.1. *The high electronic mobility transistor (HEMT)*

In JFETs, MESFETs or MOSFETs, one of the major limitations for the speed of carriers in the channel arises from the collisions between these carriers and the charged centers composed of the doping atoms. In order to eliminate this effect, it is necessary to deport these doping atoms into the gate insulator, which is then made up of another semiconductor, with a larger forbidden bandgap than that of the semiconductor channel, giving rise to a heterojunction. For III–V semiconductors, this can be achieved with near perfect continuity of the crystalline lattice at the interface ($Al_{1-x}Ga_xAs/AsGa$; $Al_{1-x}Ga_xN/GaN$). The carriers flowing into the channel then profit from increased speed while retaining the confinement due to the potential barrier of the heterojunction, which also plays the role of a gate-channel diode with a lower reverse current than in a classical MESFET. These transistors are thus faster (high-frequency applications) due to a higher mobility of carriers in the channel. In silicon MOSFET, this increased mobility can be obtained by placing the channel under mechanical strain according to the crystalline parameters of Si/Ge_xSi_{1-x} bilayers.

3.3.2.2. *Power MOSFETs and insulated grid bipolar transistor (IGBT)*

The parallel structure at the insulator/semiconductor interface of the channel in the field effect transistors described above suggests that the electric field born by the insulator in the off-state regime is proportional to $|V_{DS} - V_{GS}|$. This limits voltage V_{DS} to a value compatible with the electric field that the insulator can withstand. We can only increase V_{DSmax} at the expense of an augmented insulator thickness, which leads to a significantly deteriorated transconductance. In order to escape this limitation, a vertical architecture ("D", "V", "T" shaped, honeycomb, etc.) is implemented where the source and the gate are positioned on the top surface of the semiconducting wafer while the drain is located on the back-side surface (Figure 3.11) or buried and made accessible by means of a notch in the wafer (V-shaped structure). Channel pinching then occurs close to the top surface, although the greater part of V_{DS} is spread along all or part of the weakly doped semiconductive wafer (n– in Figure 3.11), which allows for the field to be limited in the insulator and to significantly increase the supported voltage of the component without any loss of transconductance. In this way, we achieve power MOS transistors capable of competing with bipolar power transistors built with the same semiconductor or even to greatly surpass them as far as switching speed is concerned. The parallel use of numerous simple cells such as that in Figure 3.11 is required in order to attain strong levels of current.

Another close structural component associates a bipolar transistor with a MOSFET to control the base of the bipolar transistor, which gives rise to insulated grid bipolar transistor (IGBT) (Figure 3.12). The channel of the N-MOSFET is positioned at the same place as above and allows, when it is in inversion regime in order to make the channel conductive, to inject electrons into the emitter and thus make the base emitter junction conductive near the surface of the semiconductor. This results in the injection of holes by the emitter into the base (n– in Figure 3.12). One fundamental difference compared with the power MOSFET is composed of a p+ rather than n+ doping zone on the back side, which here is termed the collector. Consequently, the base-collector voltage is supported by the depletion zone, which spreads into the weakly doped n-type part (n– in Figure 3.12) from the junction with the collector while in contrast the MOSFET off-state is obtained by an extension of the depletion zone from the channel. Another difference is found in the injection of holes (minority carriers) into the n– zone from the IGBT collector when it is conductive. This leads to an increase in the conductivity of this zone and thus a significantly weaker voltage drop in on-state than in MOSFET. This represents a major advantage under switching regime. The reliable operation of this type of component, nonetheless, requires the fine-tuning of its architecture in order to avoid the latch-up effect under conductive conditions, which initially appeared in the npnp structure that we find between the emitter contact (which is also the source of the MOSFET) and the collector. The switching speed remains more limited than that of MOSFET due to the need to evacuate or recombine the minority carriers that are injected from heavily doped zones. IGBTs can now be made with SiC, a semiconductor with a larger forbidden bandgap than Si, which can withstand greater electric fields and temperatures, allowing for voltages of several kilovolts to be reached in off-state regime.

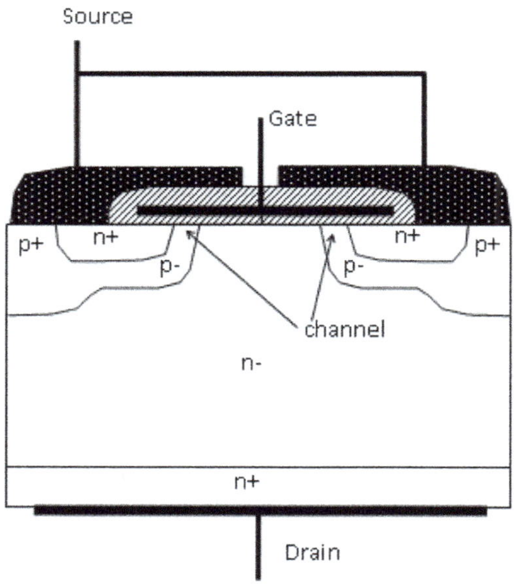

Figure 3.11. *Power N-MOSFET built in a T structure (the dark parts are metallic and the cross-hatched area insulating)*

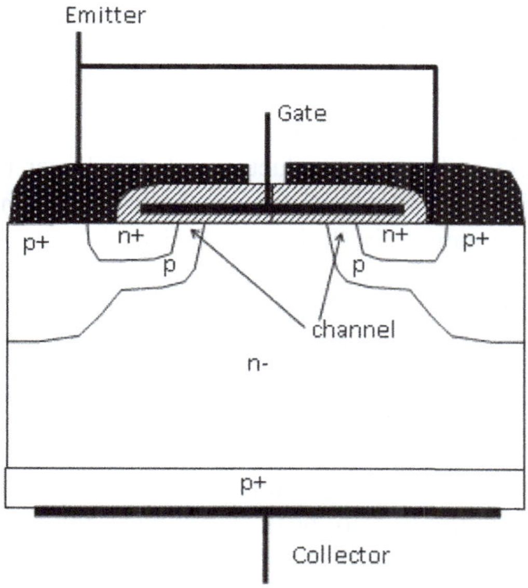

Figure 3.12. *Pnp-type IGBT*

3.4. Applications of field effect transistors

3.4.1. *Source of current and JFET biasing*

We can readily obtain a source of current I_{DSS} (zero R_S) or $I_D < I_{DSS}$ ($R_S \neq 0$) with junction FETs, as shown in Figure 3.13.

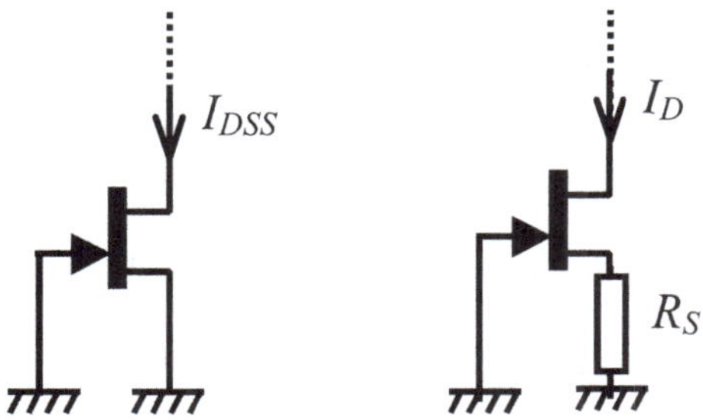

Figure 3.13. *"Automatic" polarization circuit for a JFET*

These circuits allow for the "automatic" biasing of JFETs. The use of these sources of current is widespread in integrated circuits.

3.4.2. *Amplifiers and mixers*

For small signals, FET amplifiers have the same properties as those of bipolar transistors, with a gate-source resistance that we can in most cases consider as infinite and a transconductance that is lower (10–100 times, at comparable levels). Amplifier circuits are built with common source and sometimes with common drain (follower circuit).

For strong signals, the parabolic characteristic $I_{DSat}(V_{GS})$ can be used advantageously in order to achieve a multiplication of signals, a doubling of frequency or, more often, a frequency shift (mixer), as used in hertzian signal receivers or heterodynes. Indeed, for heterodynes we need only to make the sum of the signal issued by the receiver's preamplifier, which is of low amplitude relative to V_{GSoff}, and of frequency f_1, and of the local oscillator signal, with peak to peak amplitude similar to V_{GSoff} and of frequency f_0, and finally to apply the total signal on the gate. There are also double gate transistors that allow us to perform this operation more easily. With $V_{GS} = V_{GS0} + \Delta V_0 \cos(2\pi f_0\, t) + \Delta V_1 \cos(2\pi f_1\, t)$, and writing $\alpha = \frac{V_{GS0}}{V_{GSoff}}$, $\beta = \frac{\Delta V_0}{V_{GSoff}}$, $\gamma = \frac{\Delta V_1}{V_{GSoff}}$, the JFET drain current in the parabolic approximation becomes:

$$I_D(t) = I_{DSS}\left[1 - \alpha - \beta\cos(2\pi f_0 t) - \gamma\cos(2\pi f_1 t)\right]^2$$

or by developing

$$\begin{aligned} I_D(t) = I_{DSS}[(1-\alpha)^2 + 2\beta(\alpha-1)\cos(2\pi f_0 t) + 2\gamma(\alpha-1)\cos(2\pi f_1 t) \\ + \beta^2\cos^2(2\pi f_0 t) + \gamma^2\cos^2(2\pi f_1 t)] \\ + I_{DSS}\left[2\beta\gamma\cos(2\pi f_0 t)\cos(2\pi f_1 t)\right] \end{aligned}$$

The first term is a continuous signal to which we add the constant terms derived from the development of the squared cosines, the second and third terms have the initial frequencies f_0 and f_1, the fourth and fifth have the doubled initial frequencies $2f_0$ and $2f_1$, and the last term can be rewritten as $\beta\gamma I_{DSS}\left[\cos\left(2\pi(f_0+f_1)t\right)+\cos\left(2\pi(f_0-f_1)t\right)\right]$.

It contains a frequency equal to the difference of the initial frequencies, which allows for the frequency shift. It consists of making current $I_D(t)$ enter a load impedance built with a circuit having a maximum impedance modulus $|Z_{max}|$ at frequency $f_0 - f_1$ and much lower at all the other frequencies contained in the current. In this way, we simultaneously obtain the frequency translation and the amplification with a conversion slope equal to the derivative of $I_D(t)$ relative to the variation in voltage ΔV_1, that is $g_C = \beta \frac{I_{DSS}}{V_{GSoff}}$. The output voltage is thus simply $g_C\,|Z_{max}|$ at the desired frequency, that is $f_0 - f_1$, if the output conductance of the transistor is low relative to $|Z_{max}|$. To maximize g_C, we must provide β its maximum value, which is obtained with $\Delta V_0 = V_{GSoff}/2$, corresponding to all the excursion of V_{GS} available on the parabolic characteristic. This function can also be performed with a MOSFET.

3.4.3. *Variable resistance controlled by the gate-source voltage and JFET analog switch*

For JFETs, small signal resistance r_{DS} around $V_{DS} = 0$ is given by the reverse of g_{DS} at $V_{DS} = 0$ (or also g_m), so:

$$r_{DS} = \frac{V_P}{3I_P}\frac{1}{1-\left(\frac{|V_{GS}|+\Phi}{V_P}\right)^{1/2}}$$

This resistance varies according to V_{GS} from $r_{DSon} = \frac{V_P}{3I_P}\frac{1}{\left(1-\sqrt{\Phi/V_P}\right)}$ at $V_{GS} = 0$ to infinity for $|V_{GSoff}| = V_P - \Phi$.

By switching V_{GS} between these two values, we can obtain an "on" state and an "off" state allowing for an analog switch that accepts both current directions. In the following circuits, diode D_1, positioned end-to-end with the gate-channel diode of JFET Q_1, allows us to maintain V_{GS} close to zero when Q_1 must be on, for a wide range of values for voltage V_1.

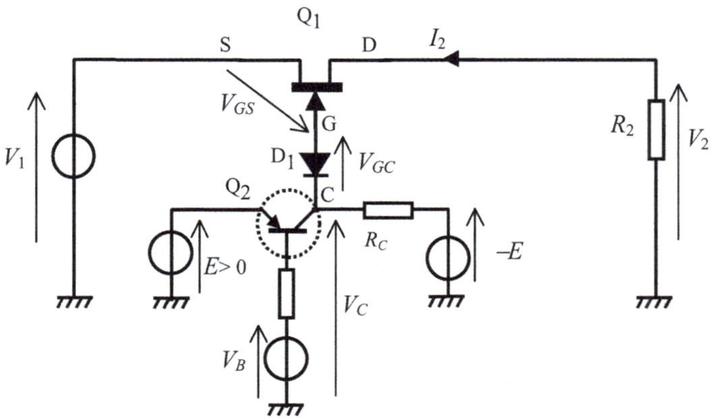

Figure 3.14. *Analog switch built with JFET, BJT and diode*

When Q_2 is in off-state, $V_C = -E$, diode D_1 bears zero voltage drop since the gate current of the JFET is negligible due to the fact that the two diodes are positioned end-to-end; Q_1 then remains off so long as $V_1 > -E + |V_{Gsoff}|$.

When Q_2 is saturated, $V_C = E$ and diode D_1 will be off if $V_{GC} < 0$, which then leaves the gate-channel diode the possibility of being under zero voltage, or $V_{GS} = 0$, inducing a conductive state of Q_1 with resistance r_{DSon}.

3.4.4. *Switching circuits and CMOS elementary logic gates (with complementary MOSFET)*

The analog switch based on MOSFETs is viable on condition that on-state resistance is not too dependent on the applied voltage. With the insulated gate of the MOSFET, the technique applied to JFET cannot be used and the problem has to be resolved by combining two complementary enhancement NMOS and PMOS transistors complementary MOS (CMOS). Both must be set in the conductive state (the PMOS by $V_{G1} = -E < -|V_T|$; the NMOS by $V_{G2} = E > |V_T|$) or in off-state (the PMOS by $V_{G1} = E$; the NMOS by $V_{G2} = -E$) in Figure 3.15. In the conductive state, if V_e increases from zero, the NMOS will become less conductive ($V_{G2} - V_e$ decreasing) but this will be compensated by the PMOS that becomes more conductive ($V_{G1} - V_e$ becoming more negative). The roles are inversed for $V_e < 0$, but the resulting effect on the overall conductance of the switch is identical.

On the one hand, using this basic circuit (or the JFET one) we can create simple switches, simple or double inverters and multiplexers, which are essential to commute logical or analog signals to or from distinct circuits, and also to allow for on demand insulation or contact of distinct circuits.

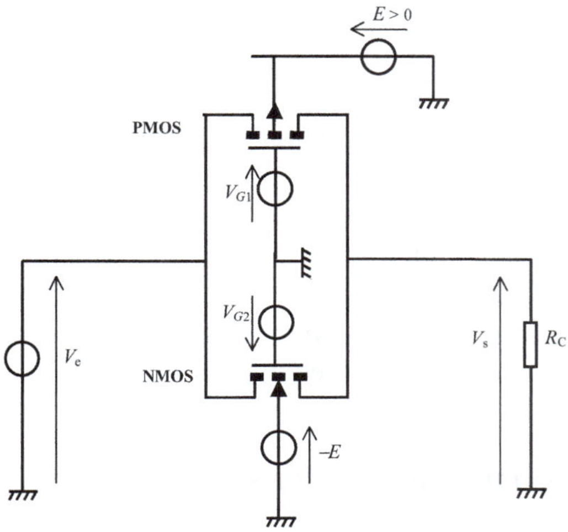

Figure 3.15. *Complementary MOSFET analog or logical switch (CMOS)*

On the other hand, the circuits of the elementary combinatorial logic are made by assembling NMOS and PMOS transistors.

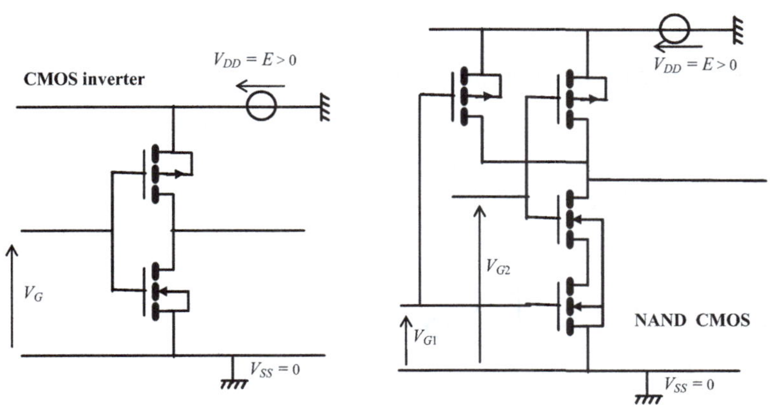

Figure 3.16. *Basic combinatorial logic functions made of complementary MOS transistors (CMOS)*

EXERCISE.– Establish the truth tables of the two circuits in Figure 3.16.

These circuits form the basis for the considerable development of logic and digital electronics (programmable logical circuits, microprocessors, microcontrollers, DSP, etc.) in CMOS technology since the 1970s.

While the logical functions may also be performed with bipolar transistors (TTL (transistor-transistor logic) and ECL (emitter-coupled logic) technologies), only the CMOS technology will be presented here since it alone is still used in current developments. Moreover, the switching and multiplexing functions are of course based on field effect transistors, which are the only ones that can be traversed by current in both directions (and for this reason occasionally referred to as bipolar, which may entail some confusion of terms when applied to the bipolar transistor, which is a reference to the presence of both types of charge carriers rather than both current directions!) and to offer an equivalent circuit that is purely resistive under the non-pinched-off channel regime.

3.4.5. *Combinatorial logic functions*

The high and low levels of voltage in these circuits are noted indifferently as H and L or 1 and 0, as in all of the following logic functions. Only the individual operating functions will be presented here, while the synthesis of complex functions concerning industrial computing methods is discussed elsewhere in specialized textbooks. The symbols used here are those that predated the normalized symbols imposed by recent norms but still currently in use now.

3.4.5.1. *Elementary combinatory circuits*

Using the circuits in section 3.4.4, we can build the following circuits easily, with the complement and invert functions indicated by a small ring on the input or output line, and by a bar above the logical variable:

– *NO (inverter):*

a	S
L	H
H	L

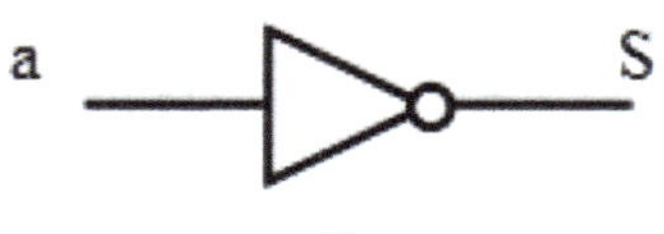

$S = \bar{a}$

The "buffer" is a function that simply copies the input variable to the output (S = a) and it is symbolized by a triangle without inverter so as to profit from a higher output current.

– *AND:*

b	a	S
L	L	L
L	H	L
H	L	L
H	H	H

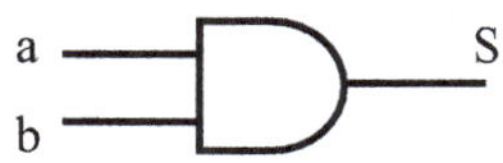

$S = a * b$

– *OR:*

b	a	S
L	L	L
L	H	H
H	L	H
H	H	H

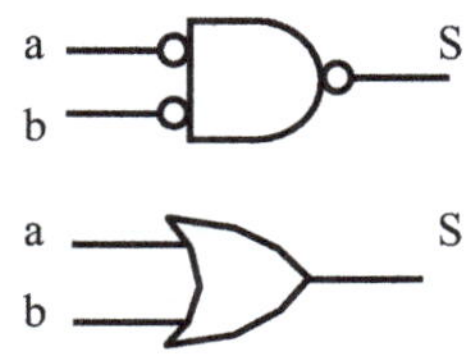

$S = a + b$

– *NAND:*

B	a	S
L	L	H
L	H	H
H	L	H
H	H	L

$S = \overline{a * b}$

– *NOR:*

B	a	S
L	L	H
L	H	L
H	L	L
H	H	L

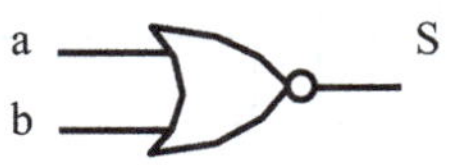

$S = \overline{a + b}$

EXERCISE.– Draw the circuit of the NOR gate from that of the NAND gate in Figure 3.16 by inverting the serialization and parallel implementation of each type of MOSFET. Check the truth table.

– *XOR*:

b	a	S
L	L	L
L	H	H
H	L	H
H	H	L

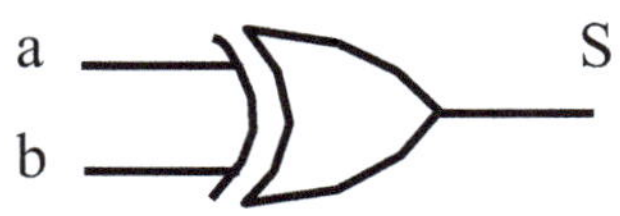

$S = a \oplus b$

The XOR comprises the basic circuit for the addition of natural binary numbers. Indeed, if we use 1 and 0 rather than H and L, we obtain for the result S the unit figure of numbers comprising 2 bits since: 0 + 0 = 00; 0 + 1 = 1 + 0 = 01; 1 + 1 = 10. To make the full 2 bit half-adder, we must simply calculate the carry, which is of course equal to 1 only in cases where the 2 input bits are at 1. So the calculation of the carry is done with an AND circuit.

$R = a \times b$; $S = a \oplus b$

b	a	R	S
0	0	0	0
0	1	0	1
1	0	0	1
1	1	1	0

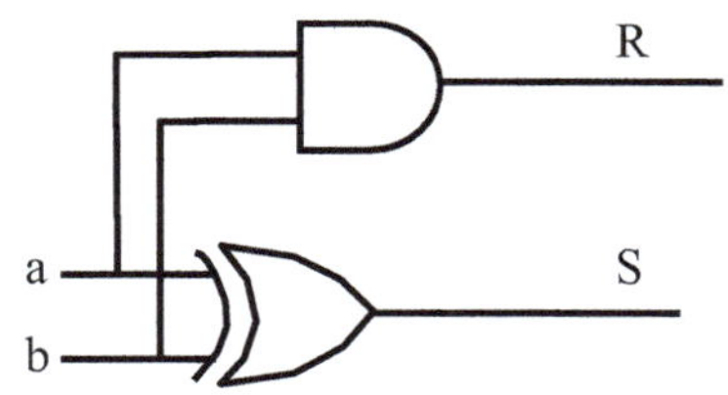

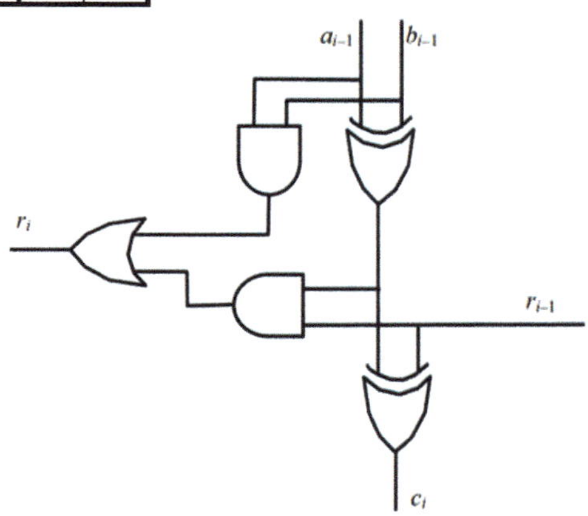

Figure 3.17. *Simple binary half-adder without entering remainder at top and full adder with an input remainder or borrow at bottom. Note that the direction of data adopted in the latter circuitry mimics the hand-written addition, where figures are written down with increasing weight from the right to the left*

This circuit is called a half-adder (top of Figure 3.17) since it can only be used for the lowest bits (units) of a natural binary number coded with n bits. Indeed, it does not account for an input remainder (or borrow) necessary if the weight of the specified bit is greater than 2^0.

If A and B are the decimal equivalents of two numbers coded in natural binary system, whose bit weight 2^i are, respectively, a_i and b_i, only taking values 0 or 1:

$$A = a_{n-1}\times 2^{n-1} + a_{n-2}\times 2^{n-2} + \ldots + a_i\times 2^i \ldots + a_1\times 2 + a_0$$

and

$$B = b_{n-1}\times 2^{n-1} + b_{n-2}\times 2^{n-2} + \ldots + b_i\times 2^i \ldots + b_1\times 2 + b_0$$

As a general rule, if $C = A + B$, we must then calculate for each bit of rank i the digit c_i resulting from the sum $A + B$ for the bit of order i accounting for the input remainder (or borrow) r_{i-1} that results from the addition of lower order and remainder (or carry) r_i that will act as input remainder for the addition of higher order. This corresponds to the arithmetic calculation $a_i + b_i + r_{i-1} = 2\times r_i + c_i$ where the term including factor 2 symbolizes the weight increase of one unit, which is necessary for the output remainder (or carry), with the + sign symbolizing addition under binary system. The truth table is that of the addition of 3 bits, implemented by the circuitry of the full adder shown at the bottom of Figure 3.17.

a_i	b_i	r_{i1}	r_i	c_i
0	0	0	0	0
1	0	0	0	1
0	1	0	0	1
1	1	0	1	0
0	0	1	0	1
1	0	1	1	0
0	1	1	1	0
1	1	1	1	1

We deduce using Boolean logic methods that the logical expression is the following (where the + sign represents a logical *OR while* ⊕ is the XOR operator):

$$c_i = (a_i \oplus b_i) \oplus r_{i-1}$$

$$r_i = a_i * b_i + a_i * r_{i-1} + b_i * r_{i-1} \quad \text{or rather:}$$

$$r_i = a_i * b_i + r_{i-1} * (a_i \oplus b_i).$$

In the full-adder, the flow of remainders is directed toward the left, like in doing the handwritten addition of numbers, in detail above (bottom schematics in Figure 3.17) and as a unique symbol below (Figure 3.18).

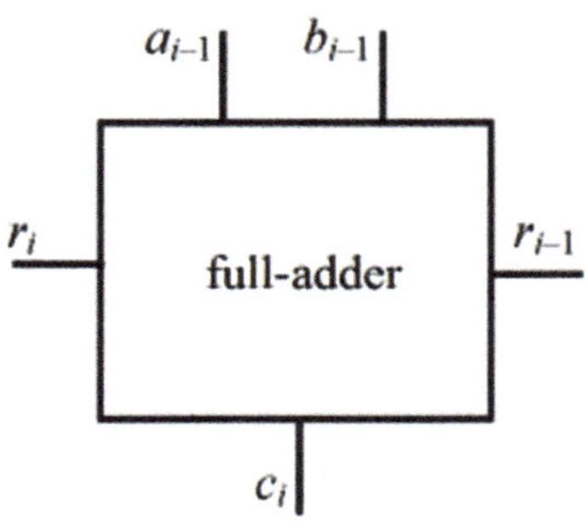

Figure 3.18. *Full adder circuit*

The architecture possible for the addition of two binary numbers is shown in Figure 3.19.

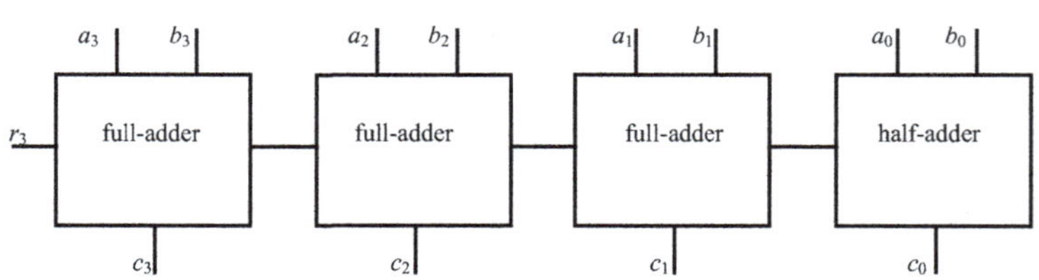

Figure 3.19. *Full-adder of two natural 4 bits binary numbers*

It should be noted that the result's bit number is increased by one unit since the remainder *r*3 can take value 0 or 1, and consequently, the result is written in the order of decreasing bit weight from left to right *r*3 *c*3 *c*2 *c*1 *c*0.

We can also perform the subtraction $C' = A - B$ by complementing inputs b_i and replacing the half-adder by a full-adder with an input remainder positioned at 1, on condition that the result is positive. If the result has to be potentially negative, we will have to resort to operations on signed numbers described below.

3.4.5.2. *Elementary arithmetic of signed numbers*

Natural binary code allows us to code non-signed numbers, taking equivalent decimal values from zero to $2^n - 1$ (full scale) in the case of a number coded with n bits. Among the binary codes that let us represent numbers with a sign (or signed numbers), we will only consider those in which the representation of zero is unique.

1) One of these is representation by 2's complement, where the positive numbers up to $2^{n-1} - 1$ are represented by the $n - 1$ bits for the least weights and zero for the most significant bit (MSB), and the negative symmetrical numbers by 2's complement of the positive number, including the MSB that is then 1 for negative numbers (= signed bit). We obtain the opposite of a positive number by changing all of the bits from 0 to 1 or from 1 to 0, then by adding 1 to the least significant bit (LSB). Symmetrically, we obtain the absolute value of a negative number (MSB at 1) by subtracting 1 at the LSB of its representation, then by reversing all the bits. The addition of two numbers then works correctly, whether they are positive or negative, and subtraction is replaced by the calculation of the opposite, then the addition. This system is well adapted for cases where we need to perform operations such as addition and subtraction, and even multiplication and division, which require shifts and comparisons (see below), for example in the case of the circuits called "arithmetic & logic units" that form the basis of microprocessors. The drawback of this representation is that all the bits change when we go from 0 to the representation of the number -1 composed of n bits equal to 1.

2) Another representation is that with excess of half scale or equivalently by shifting all numbers by a half scale: here we add 2^{n-1} to all the numbers, that is the half-scale (= half of 2^n), such that zero becomes the median value of the complete scale, that is 2^{n-1}. The signed bit is still that of the MSB but now is worth 0 for the negative numbers and 1 for the positive ones. There is no longer any discontinuity on passing by zero, unlike in the previous solution, which is advantageous when there is a time evolution of the coded

number, as in the case of analog-digital converters. Conversely, for digital-analog converters, it is very easy to restore the zero value by simple analog shifting.

In this representation, the effective binary value of the positive numbers from 0 to $2^{n-1}-1$ is simply that of the natural binary number obtained by omitting the MSB. The negative numbers occupying the interval $[0,\ 2^{n-1}-1]$ or $[1,\ 2^{n-1}-1]$ if we want to code them with the same number of bits as for positive numbers, that is $n-1$, and the zero is represented by 2^{n-1}. In order to obtain the absolute value of a negative number N coded by the equivalent natural binary code of $N+2^{n-1}$, we have to reverse all of the bits from 0 to 1 and from 1 to 0 (complement at 1), then calculate the result of the addition in which we add the unit (that is LSB = 1). Conversely, to obtain the representation of a negative number whose absolute value is known (in natural binary code and with the MSB at 1), we must deduct 1 from this representation of the absolute value then change all the bits from 0 to 1 and from 1 to 0, including the MSB.

The addition of two numbers functions correctly in this representation, whether they are positive or negative, on the condition that we replace the MSB in the addition result by the remainder of the sum of the two MSBs, which actually corresponds to the value of the bit whose weight is 2^n. As with the representation by 2's complement, the operation must also be legitimate, that is the result of the addition of the signed numbers must remain in the interval $[-2^{n-1}\ ,\ +2^{n-1}-1]$ formulated in decimal code, and we replace subtraction by the calculation of the opposite then addition.

EXERCISE.– Process several examples of legitimate addition and subtraction of positive and negative numbers between −128 and +127 in the 8 bit binary representation with excess of 128.

The multiplication of two binary coded numbers A and B demands that we consider several aspects such as the total number of bits, the final sign and the calculation method of the resulting mantissa. If two signed numbers are coded with one on $n-1$ bits, the other on $m-1$ bits, and both in one of the representations above, the result of the multiplication is a number coded on $n-m-2$ bits, necessary to surpass $(2^{n-1}-1)\times(2^{m-1}-1)/2$. And with regard to the final sign, the sign rule is always verified by applying the XOR rule, followed by an inversion only if the negative sign is represented by the MSB = 0 (level L), since two different signs involve a negative sign and two identical signs a positive sign. The mantissa calculation of the result requires to do the product of the absolute values of A and B, and then to apply the corresponding MSB to the final sign. This method is analogous to that of handwritten multiplication, calculated step by step. If A is coded on $n-1$ bits and if $B = b_{m-2}\times 2^{m-2} + b_{m-3}\times 2^{m-3} + \ldots + b_i\times 2^i \ldots + b_1\times 2 + b_0$, we add the partial results of $A\times b_0$, $A\times b_1\times 2$, ... $A\times b_i\times 2^i$ up to $A\times b_{m-2}\times 2^{m-2}$, which are zero when the relevant bit is zero. In cases where the relevant bit b_i is equal to 1, the partial result $A\times b_i\times 2^i$ is announced by a shift in all the bits of number A by a quantity equal to i, so that the weight of all the bits of number A happens to be increased by i. The next operation consists of making the total of the partial results. And finally, if the result is negative, we must apply the changes described above to go from absolute value to the representation of the negative number. It is possible to perform such a multiplication with wired gates by using combinational functions, more easily when one of the two numbers A or B is constant.

3.4.5.3. *Binary comparator*

The comparator allows us to detect the equality of two numbers and possibly indicate the larger or the smaller.

The comparison occurs on each bit of the same weight a_i and b_i that belong to two binary numbers A and B, starting with the MSB. The truth table is as follows:

a_i	b_i	$a_i = b_i$	$a_i > b_i$	$a_i < b_i$
0	0	1	0	0
0	1	0	0	1
1	0	0	1	0
1	1	1	0	0

In fact, the elementary cell comprises a supplementary input called "enable input" (EI) that uses the detection result of higher weight. Indeed, this comparison is useful in so far as the more significant bits are equal; but if an inequality is detected, there is no more value in comparing bits of lower weights, and in this case subsequent comparisons should be stopped since they are of no use. The cascade method performs the bit equality detection sequentially starting from the MSBs. There is also a parallel method for comparing all bits simultaneously, with the result for each bit acting as a "go-no go" block for comparisons of lower weight. Regardless of the detail of the combinatorial operations used, all based on AND and OR functions in addition to inversions, these instances operating on a given number of bits can themselves be positioned in cascade in order to compare coded numbers with more bits because of the inputs and outputs taking into account or delivering the result of the instance performing the comparison for all bits of lower or higher weight.

3.4.5.4. *Multiplexers, demultiplexers, encoders, decoders, transcoders and other combinatorial functions*

The multiplexing function of 2^n binary data to a single output is a switching circuit that establishes a link between an input selected because of values placed on the n address inputs and the output, using 2^n MOS complementary transistor switches (Figure 3.15), connected on one side at each input and on the other at the output. Two applications are the function generators and the data transmission with parallel-series conversion.

Demultiplexers perform the reverse function, that is they have a single input, which is switched toward one of 2^n outputs chosen by means of values on the n address inputs. These circuits are also called decoders in so far as they act to transform numbering codes (binary to another binary code; binary to octal; binary to hexadecimal, etc.).

Encoders are circuits with 2^n data inputs and n output addresses. The inputs are arranged according to an order of priority and the outputs provide a code on n bits indicating the rank of the highest priority input. For example, for $n = 3$, we can find the next following truth table, where Di (i = 0-7) represents the data in an increasing order of priority, with the active level to low level (L) and similarly for the EI; A0, A1, A2 the outputs, GS and EO the enable outputs (X is a non-significant value). When inputs and outputs are active at the high level, everything is reversed. This function, called priority encoder, is useful in "flash" converters (see Chapter 2 of Volume 3 [MUR 17b]) and in code conversions such as those performed in keyboard coders. They allow for the position of the first change of state to be located in an ordered sequence of 2^n data, indicating this position by a coded number on n bits in natural binary code.

EI	D0	D1	D2	D3	D4	D5	D6	D7	A2	A1	A0	GS	EO
H	X	X	X	X	X	X	X	X	H	H	H	H	H
L	H	H	H	H	H	H	H	H	H	H	H	H	L
L	X	X	X	X	X	X	X	L	L	L	L	L	H
L	X	X	X	X	X	X	L	H	L	L	H	L	H
L	X	X	X	X	X	L	H	H	L	H	L	L	H
L	X	X	X	X	L	H	H	H	L	H	H	L	H
L	X	X	X	L	H	H	H	H	H	L	L	L	H
L	X	X	L	H	H	H	H	H	H	L	H	L	H
L	X	L	H	H	H	H	H	H	H	H	L	L	H
L	L	H	H	H	H	H	H	H	H	H	H	L	H

Transcoders allow us to change from one code to another, notably for the display of data. The most common instance of this consists of passing hexadecimal codes or BCD to that allowing display in seven segment characters.

More generally, by making use of elementary logical gates and switches, possibly organized in a matrix, the methods of description and logical synthesis such as those provided by the manufacturers of programmable logical circuits, allow us to synthesize almost any combinatorial function of type $S_j = f_j(E_1, \dots E_i, \dots E_n)$, with *n* input variables *E1* to *En,* and *m* output variables *(j =1 to m*). The circuit obtained has the advantage of providing each output variable with a delay limited only by the propagation and transition time of the technology used. Such a quality can be primordial in "real-time" systems implemented to rapidly perform arithmetic calculations bearing on a variable operand composed of input bits E_1 to E_n and with the assistance of another immutable operand. As we see in PLA circuits ("programmable array logic") or CPLD circuits ("complex programmable logic device"), the addition of a layer of flip-flop output sequential circuits, or in some cases input circuits, lets us escape from random delays and arrive at synchronized output updates with the clock.

3.4.5.5. *Three state output circuits*

When several lines of data are connected on a bus, they cannot function simultaneously with outputs connected in parallel without risking the creation of a conflict that could lead to the destruction of the transistors operating in conductive regime at this time. Accordingly, it is indispensable that all outputs connected on the same line be isolated with the possible exception of one, when one has to impose the high (H) or low (L) value of the bit in question. This indispensable precaution is made possible by a switch as seen earlier (Figure 3.15), connected between the functional output delivering the data and the bus line. This is referred to as the "three state output", which means that the output can take three states, that are insulated, H or L.

3.4.5.6. *Arithmetic and logical units*

This type of multifunctional circuit forms the heart of all microprocessors. By means of the elementary functions above, it allows us to make logical comparisons and elementary arithmetic operations, including multiplication and division, which also require the adders and shift registers seen below.

3.4.5.7. *Dynamic random access memory*

Rapid memory such as that associated with the functions performed by microprocessors are typically based on the circuit known as "one transistor" dynamic random access memory (T1) letting us control and read the charge stored in a C_S capacitor, which electronically represents the data, as shown below (Figure 3.20).

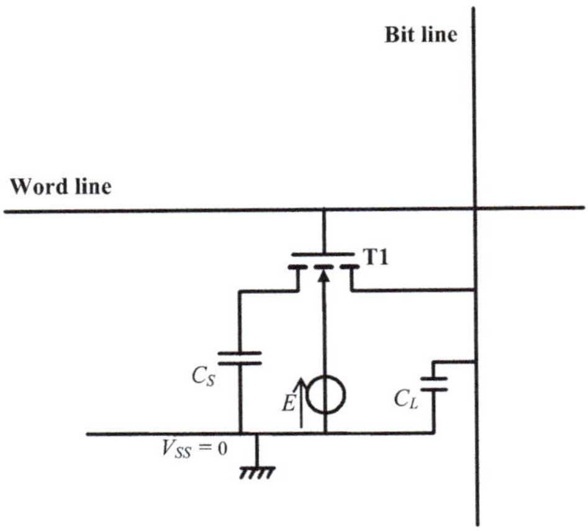

Figure 3.20. *Basic 1 transistor DRAM circuit*

Capacitor C_S acts to store either a non-zero or quasi-zero charge, featuring each state. Due to the leakage current, its charge must be periodically restored if there is no read operation. Capacitor C_L is derived from the circuit design and is inherent to the bit line's great length.

The write operation occurs by imposing potential H or L on the bit line and making T1 conductive by the potential applied on the line, greater than E (= in principle $V_{DD}/2$, that is half the power supply voltage) plus the transistor's conduction threshold.

The read operation occurs as follows (it destroys the data and must be followed by the restoration of the initial value): (1) precharge of bit line to $V_{DD}/2$; (2) control of line to make T1 conductive; (3) C_S then equalizes its potential with that of the column which varies little; (4) a detection circuit, not represented here (analog comparator, see Chapter 4), detects if C_S loses or gains charge (and accordingly voltage also) and thus determines value H or L that was stored; (5) a write operation reestablishes the charge before reading. In the absence of reading, refreshment, which is required as per a period in the order of several dozen milliseconds, occurs as for a mute read, that is without transmission of the result. Management of these operations requires other logical circuits, but ones that are common to the lines or to the columns, which makes this technology very effective in terms of occupying the surface of the silicon wafer.

Other basic circuits use three transistors and allow a non-destructive read of the data, with distinct and independent control lines and columns (Figure 3.21).

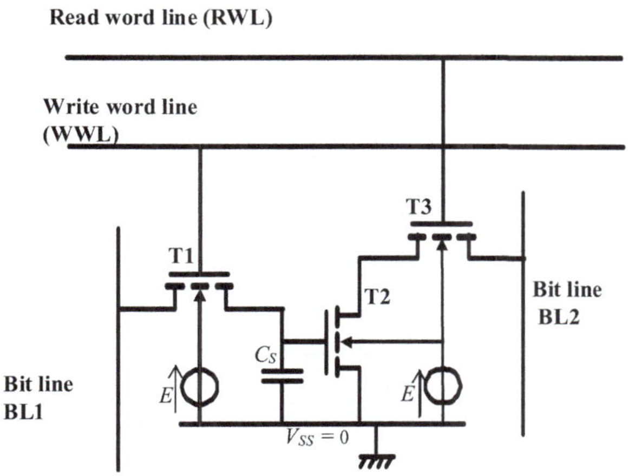

Figure 3.21. *Three transistor basic DRAM circuit*

An H state on WWL allows for the storage of data placed on BL1 in the capacitor and an L state then allows for the capacitor to be insulated again. Transistor T2 is conductive or off depending on whether the data in C_S is L or H, respectively. An impulse to state H on RWL allows for the state of T2 to be read and be transcribed in BL2. This type of memory requires more space on the wafer but far fewer operations, which can be worthwhile for fast buffer memory interfaces with the arithmetic and logical units, with storage capacity less important than in data and program memories. Without the operation, data refreshment can prove necessary but with much less frequency than for "one transistor" circuits.

3.4.6. *Sequential logic functions*

3.4.6.1. *Bistable RS flip flop*

By implementing a feedback from the output onto the input, a circuit based on two combinatorial functions becomes able to keep a state imposed earlier by one of the two inputs. Accordingly, this is a memory, also referred to as a bistable flip flop. The RS flip flop (R = reset; S = set) using two NOR gates operates with active inputs at high state and sees its output state Q copy the high state of S or the complement of the R high state, which is performed by the circuit shown in Figure 3.22 and the truth table below, where Q_{n-1} and its complement represent the preceding output states.

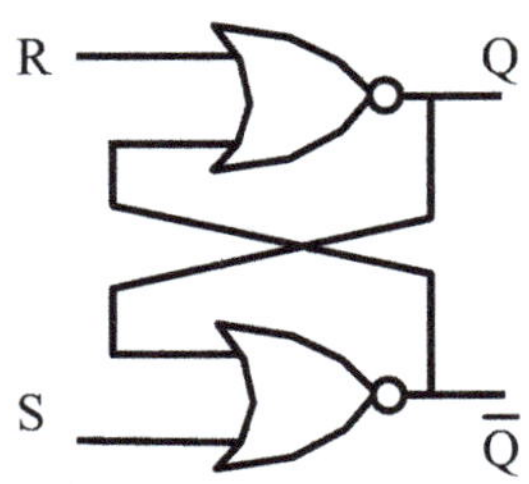

S	R	Q	$\overline{Q}$
L	L	Q_{n-1}	$\overline{Q}_{n-1}$
H	L	H	L
L	H	L	H
H	H	L	L

Figure 3.22. *RS flip flop*

The last state is clearly incoherent since both outputs are not complementary, and simultaneous H states must be forbidden for R and S. The output change occurs when R or S shifts to state H and this operation is called asynchronous since it is not coordinated with a clock signal. We will represent the RS flip flop with the active inputs at level H as in Figure 3.23.

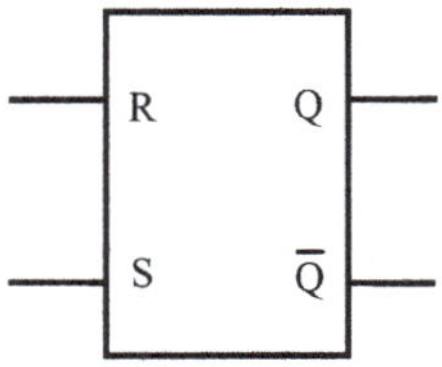

Figure 3.23. *Bistable RS flip flop*

A similar circuit can be made with two NAND gates, but the inputs are then active at level L.

EXERCISE.– Study the RS flip flop wired with two NAND gates and establish its truth table.

3.4.6.2. *Static CMOS memory*

We use the memorization property of the RS flip flop. This requires simply adding two transistors working as a switch to insulate this memory cell from the lines and columns that convey the data or to perform the writing or reading operations.

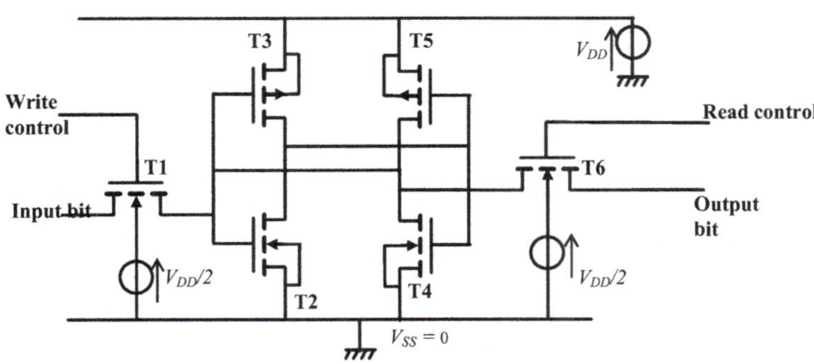

Figure 3.24. *Static CMOS memory cell (SRAM) (note: only the T lines are connections, crossings are not connections)*

This requires six transistors but retains the stored data indefinitely so long as the circuit's continuous power supply is maintained.

3.4.6.3. *RS flip flop with validation (RST) and D flip flop*

To induce a change at outputs only when enable is provided by the high state of input T, we make the circuit presented in Figure 3.25.

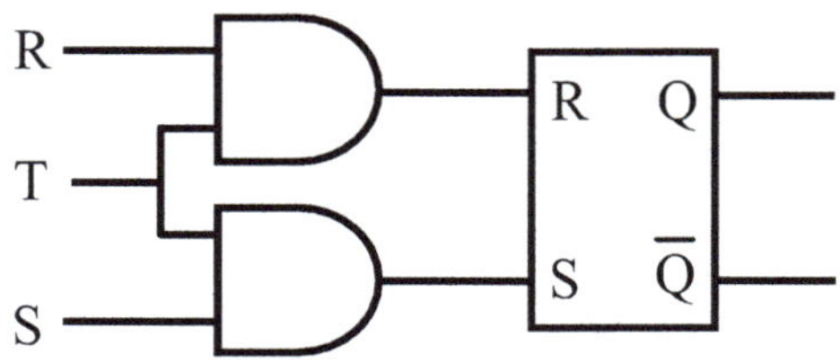

Figure 3.25. *RST flip flop*

But the state R = S = H always provokes the same incoherence in terms of outputs. To avoid this, we must oblige S to always be complementary to R, which becomes the unique data D. Hence, the circuit of the D flip flop in which input T determines the state of outputs when it is in the high state and leaves them in the state in which they are when it is in the low state.

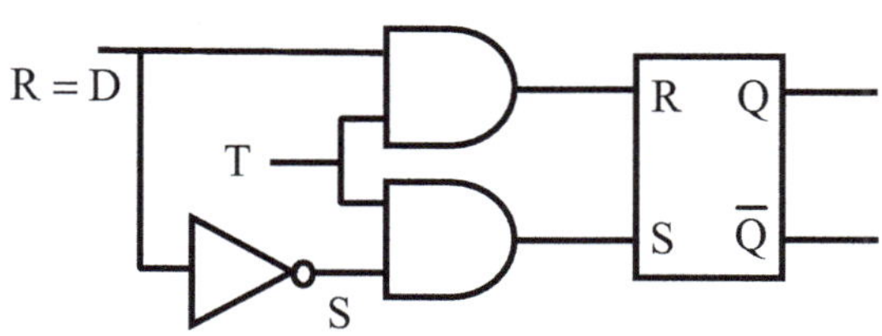

Figure 3.26. *D latch with validation (T)*

The D latch is represented in Figure 3.26, where T is the validation input authorizing the copy of D on Q when T is at the high level.

One example of the variable logic chronograms of this latch is given in Figure 3.27. However, if there is a change in D while T is still at state H, output Q follows D, as for the second high state of D in the figure.

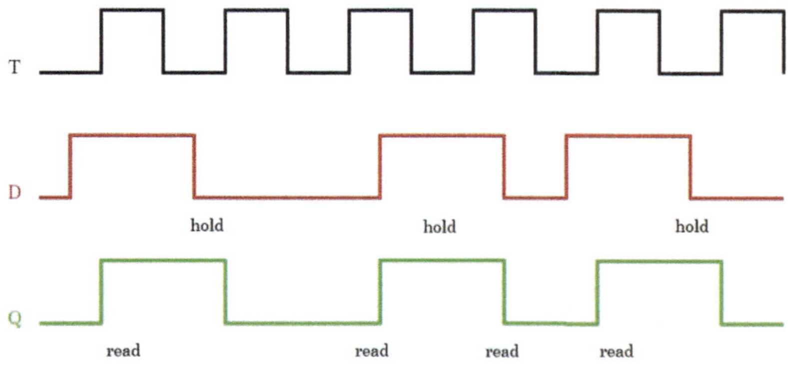

Figure 3.27. *Chronogram of the D latch activated by state H of T*

To ensure full synchronism with a clock, we must replace these flip flops by edge triggered flip flops of the master-slave variety, as described in the following.

3.4.6.4. *Synchronous D edge-triggered flip flop*

We use two D latches with validation on the high state in cascade so that when the first one (the master) copies the input data D to the first output Q' while the second one (the slave) is idle, the clock subsequently triggers the copying of the intermediate state Q' on the second output. Indeed, if clock h is at state L, the first flip flop copies D on Q' while the second keeps the previous state on its output Q. This transition indeed occurs on the rising edge of h = Ck ("clock"). If we use D latches manufactured with NAND gates, the validation occurs on the state L and triggering occurs on the falling edge.

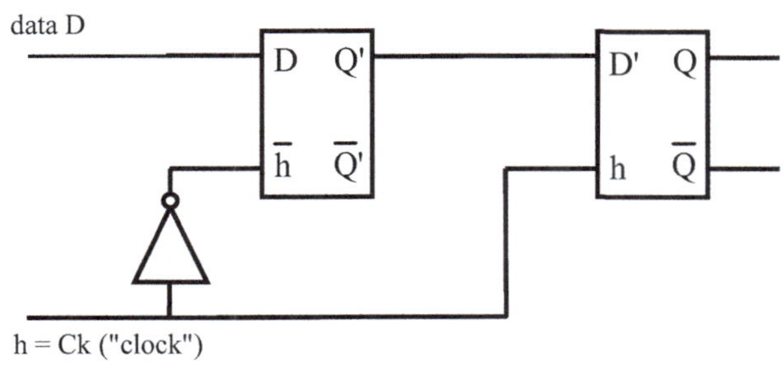

Figure 3.28. *Synchronous D latch on rising edge*

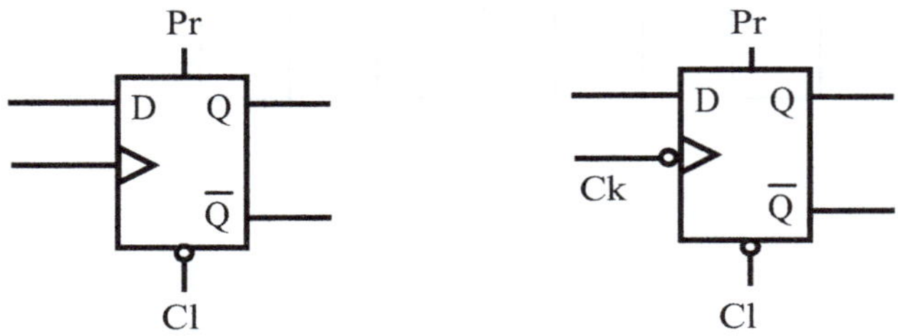

Figure 3.29. *Triggering D latch on rising edge to the left, and on falling edge to the right, with PRESET input active at H and CLEAR active at L*

The detailled circuit made with logical gates (typically NAND) in practice varies somewhat and allows for the addition of forced inputs (Figure 3.29), which act independently of the clock to put Q either at H (PRESET input or Pr), or at L (CLEAR or Cl), by an active state either H or L, respectively.

Along the same lines, we can make master-slave RS, generally called RSh, triggered by the clock edge (h).

3.4.6.5. *Master slave JK flip flop*

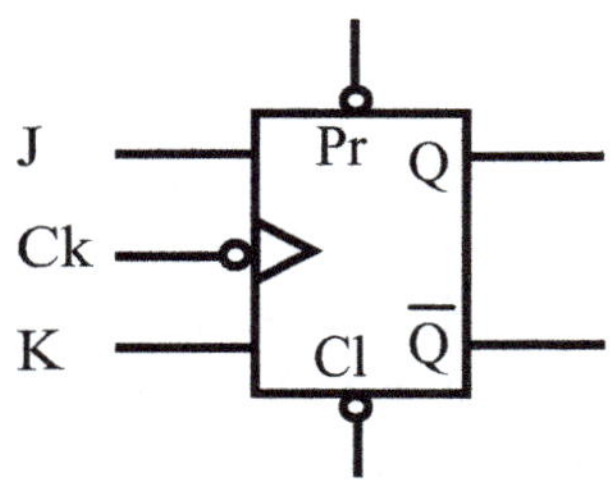

Figure 3.30. *Downwards edge triggered JK flip flop with forcing inputs active at level L*

By adding a combinatorial circuit to differentiate two distinct input data (or more), we use the JK flip flop in which the input values J and K determine the operation. The flip flop occurs on one of the clock edges, typically falling, due to the master-slave structure similar to that of the previous flip flops:

– if J = K = L, outputs Q and $\overline{Q}$ do not change;

– if J = H and K = L, Q shifts to H (or stays);

– if J = L and K = H, Q shifts to L (or stays);

– if J = H and K = H, Q necessarily changes (TOGGLE mode).

Operations are summarized in the truth table below in which Q_{n-1} and $\overline{Q}_{n-1}$ are the output states at the previous transition (if $Q_n = Q_{n-1}$, there is no change) and X any state. The forced inputs (active at level L in Figure 3.30 and in the truth table) always act immediately, independently of the clock. Since the forced inputs are of type R and S, both simultaneous active states must be disallowed.

Ck	J	K	Pr	Cl	Q	$\overline{Q}$
X	X	X	L	H	H	L
X	X	X	H	L	L	H
X	X	X	H	H	Q_{n-1}	$\overline{Q}_{n-1}$
↓	L	L	H	H	Q_{n-1}	$\overline{Q}_{n-1}$
↓	H	L	H	H	H	L
↓	L	H	H	H	L	H
↓	H	H	H	H	$\overline{Q}_{n-1}$	Q_{n-1}

3.4.6.6. *Counters*

The basic binary counter function is implemented by looping the output $\overline{Q}$ on input D of a synchronous D latch (Figure 3.31). Since the output only changes on one of the two edges, rising in the example below, the period of signal Q is doubled relative to that of the clock.

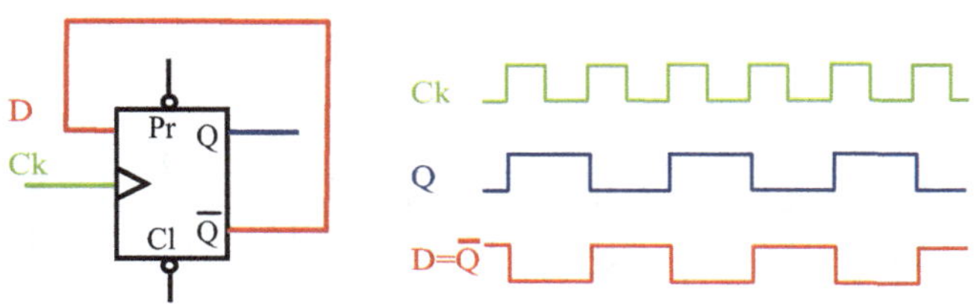

Figure 3.31. *One stage binary counter or frequency divider by 2, with its chronogram*

If we suppose $Q = L$ and $\overline{Q} = H = D$ at the moment of the clock's first rising edge, this would entail the shift of Q to H and $\overline{Q}$ to L. The following change would only occur with the new rising edge of the clock. Thus, there is division of the frequency by 2.

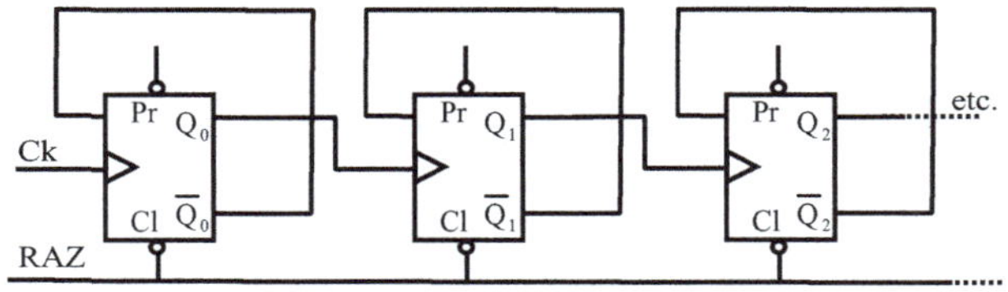

Figure 3.32. *Asynchronous binary counter*

The asynchronous binary counter is derived from the divider by 2 by cascading n dividers by 2 as in Figure 3.32 by using the forced CLEAR to initialize all the outputs Q to L state.

Next, the H level spreads with half an occurrence on each stage, which leads to a state q_0, q_1, … q_{n-1} of the n outputs Q0, Q1, … Qn–1, so that the decimal equivalent of the n number is $n = q_0 + q_1{\times}2 + q_2{\times}2^2 + \ldots + q_{n-2}{\times}2^{n-2} + q_{n-1}{\times}2^{n-1}$, where we have assimilated the number 0 to L and the number 1 to H for each bit. The outputs state thus exactly reflects the equivalent natural binary of number n. Since the clock of each stage is taken at output Q of the previous stage, propagation delays add to each other and we must wait until the last level has changed before reading all of the Q outputs. The count occurs modulo n since when all the outputs are at "1", the next state corresponds to all the outputs at "0" and the cycle begins again.

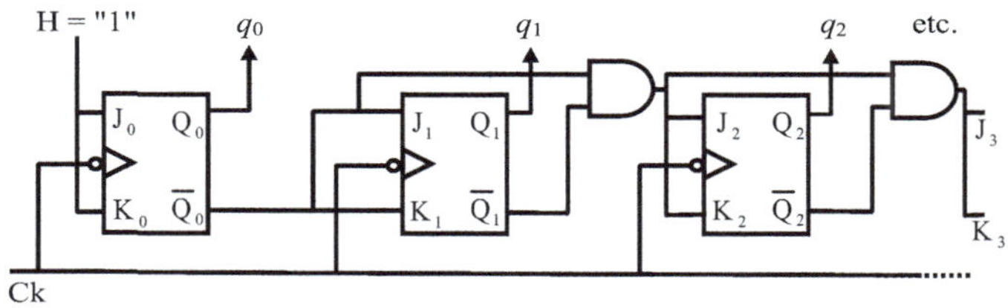

Figure 3.33. *Synchronous binary counter*

Conversely, the synchronous count becomes possible if all of the flip flops are controlled by the same clock. So there is a need to dissociate the data that propagate and the single clock, which is feasible with JK flip flops. In such cases, we systematically use the "TOGGLE" J = K = H mode for the state change at each triggering edge, but permitted only if all the previous stages had also changed their output at the previous clock strike. This gives the circuit of Figure 3.33 where we have omitted the forced input, but the CLEAR input still allows for reset.

EXERCISE.– Determine the successive state of all the outputs from the synchronous counter by establishing a table or a chronogram and assuming all outputs q_0, q_1, q_2 ... initially as zero.

Synchronism of output changes in all stages enables two more complex but extremely useful modes of operation for electronic systems.

First, those in which a determined output combination is detected by a combinatorial circuit in order to reset them to zero. This lets us design 10 modulo, 16 modulo or any integer number modulo counters. Indeed, counting in the decimal system can be performed by cascading 10 modulo counters in asynchronous or synchronous mode. This system also lets us design frequency dividers for any integer number.

Second, another application takes into account an input that determines whether an up-count or a down-count is performed, by means of the control of J and K by an XOR between the complement of the up-count/down-count bit and each output Q0, Q1, Q2 ... prior to sending it into each AND gate of the synchronous counter. The forced inputs PRESET and CLEAR of each flip flop are then used to initialize (or load) the counter to a value between 0 and $2^n - 1$ represented by its natural binary equivalent, then either up-counting or down-counting from this value. In the second case, we obtain a digital timer by detecting the zero-crossing of all the outputs simultaneously. The timing accuracy is subject to that of the clock, which can be very good. This type of counter is called "prepositioning or pre-loading counter". We can easily design the combinatorial circuit which, for each stage, forces Q*i* to a level H or L from the same level H or L of the prepositioning logic variable E*i* by means of an inverter and two AND or NAND gates according to the active type of the forced inputs and on the LOAD variable (example in Figure 3.34 with LOAD active at level H).

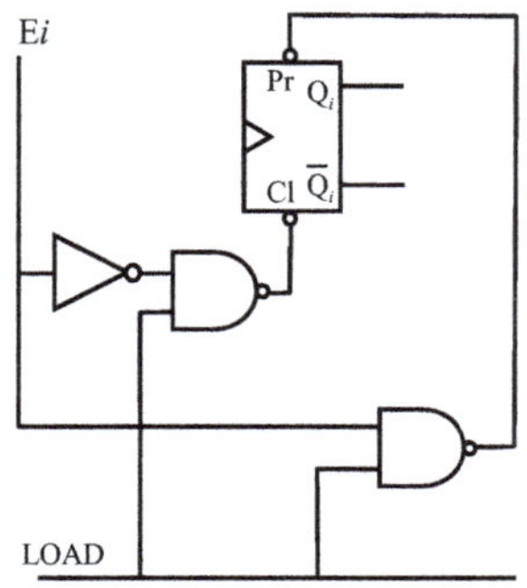

Figure 3.34. *Prepositioning circuit of a counter stage (or shift register)*

3.4.6.7. *Shift registers*

Unlike counters, shift registers propagate a binary word made by the state of each output Qi from stage to stage, with each consisting of a synchronous D latch or JK flip flop. This binary word can be introduced either in series mode after a certain number of clock strikes, or in parallel mode by means of the same technique as that used in the prepositionable counter with the loading circuit (Figure 3.34). The register acts like a memory whose contents can be translated and then retrieved, either in serial mode or in parallel mode. The operational mode of the register is always synchronous. The bidirectional translation of the binary word is possible with a simple combinatorial circuit, which on each stage, allows to choose whether Q$i-1$ or Q$i+1$ is applied to Di (in the case of a shift register composed of synchronous D latches).

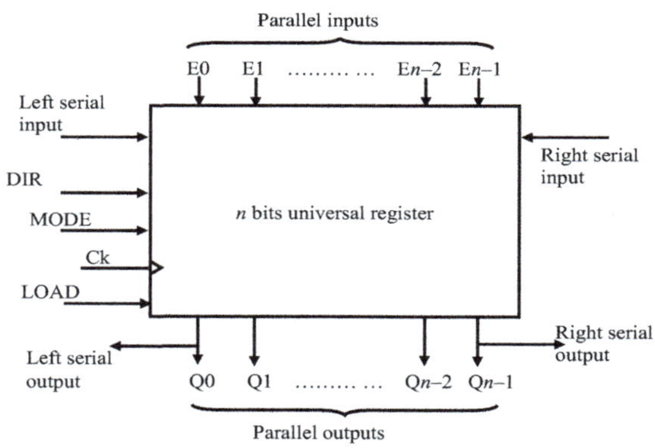

Figure 3.35. *Universal shift register*

We can thus select operations of parallel or serial load and bidirectionality or not. Serial or parallel reading is still possible since it is performed on the output of the last flip flop or on all of the outputs, respectively. "Universal" registers offer all of these functions, with the MODE variable allowing either parallel loading or shifting.

The applications of shift registers are numerous:

– temporary memorization of information in the form of a binary word;

– parallel–series or series–parallel conversion of binary words;

– digital delay line;

– multiplication and division by 2^k of a binary number (which is reduced to a shift towards the left or the right);

– digital filtering of the "pipe-line" variety and other similar techniques;

– generation of predetermined (for the control of a process for example) and pseudo-random (noise generation) sequences.

3.5. Exercises

3.5.1. *Parabolic approximation of the JFET characteristic $I_D(V_{GS})$ in pinched-off (or saturated) regime*

For a JFET, calculate the current and pinch-off voltage I_P and V_P from the following data:

$$Z = 20\ \mu\text{m};\ L = 2\ \mu\text{m};\ a = 1\ \mu\text{m};\ \mu_n = 800\ \text{cm}^2/\text{Vs};\ N_D = 10^{16}\ \text{cm}^{-3};$$

$$\varepsilon = \varepsilon_r\ \varepsilon_0 = 11.7 \times 8.85\ 10^{-14}\ \text{F/cm};\ \Phi = 0.7\ \text{V}.$$

From this, deduce V_{GSoff}, I_{DSS} and the maximum transconductance g_{mmax}. Plot characteristic $I_D(V_{GS})$ in pinched-off (or saturated) regime from the approximation

$$I_D(V_{GS}) = I_{DSS}\left[1 - \frac{V_{GS}}{V_{GSoff}}\right]^2.$$

Determine resistance R_S to be placed in series with the source so that $V_{GS} = V_{GSoff}/2$ in the circuit drawn in Figure 3.13 when $e_g = 0$.

– Mixer operation (or 2 quadrant multiplier).

A decoupling capacitor must be set in parallel on the resistance R_S.

Calculate $I_D(t)$ when $e_g(t) = (V_{GSoff}/2)\cos(2\pi f_1 t) + V_2\cos(2\pi f_2 t)$ with $V_2 << V_{GSoff}/2$.

Having developed this, indicate the frequencies of the various terms. Numeric application with: $f_1 = 100$ MHz; $f_2 = 89.3$ MHz (FM radio band).

Calculate the conversion slope $\dfrac{\Delta I_D}{V_2}$ where ΔI_D is the variation of the term which has the frequency $f_1 - f_2$.

Answer:

We have $I_P = (e^2\mu_n Za^3 N_D^2)/(6\varepsilon L) = 3.3$ mA and

$V_P = (ea^2 N_D)/(2\varepsilon) = 7.7$ V.

Hence, $|V_{GSoff}| = V_P - \Phi = 7$ V and

$$I_{DSS} = I_P\left[1 - 3\frac{\Phi}{V_P} + 2\left(\frac{\Phi}{V_P}\right)^{3/2}\right] = 2.58 \text{ mA}.$$

So $V_{Gsoff} = -V_P + \Phi = -7$ V and $g_{m\max} = \dfrac{3I_P}{V_P} = 1.3$ mA/V or 1.3 mS.

Plotted $I_D(V_{GS}) = I_{DSS}\left[1 - \dfrac{V_{GS}}{V_{GSoff}}\right]^2$ in continuous line and exact current in dotted line:

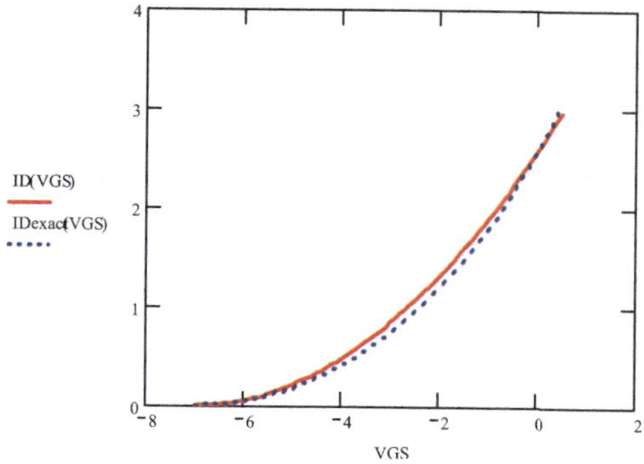

For $V_{GS} = V_{GSoff}/2 = -3.5$ V, $I_{D0} = 0.645$ mA. Since $V_{GS} = -R_S I_D$, $R_S = 5.4$ kΩ.

By placing a decoupling capacitor in parallel on the resistance R_S, we find $e_g(t) = v_{GS}(t)$ as a dynamic variation superimposed to $V_{GS0} = V_{GSoff}/2$. So:

$I_D(t) =$

$$I_{DSS}\left[1-\frac{V_{GS0}+e_g}{V_{GSoff}}\right]^2 = I_{DSS}\left\{1-\left[\frac{1}{2}+\frac{1}{2}\cos(2\pi f_1 t)+\frac{V_2}{|V_{GSoff}|}\cos(2\pi f_2 t)\right]\right\}^2$$

Write $\alpha = \dfrac{2V_2}{|V_{GSoff}|} << 1$. Then:

$$I_D(t) = \frac{I_{DSS}}{4}\left\{1+\cos(2\pi f_1 t)+\alpha\cos(2\pi f_2 t)\right\}^2$$

$$= \frac{I_{DSS}}{4}\left\{1-2\cos(2\pi f_1 t)-2\alpha\cos(2\pi f_2 t) + \frac{1+\cos 4(2\pi f_1 t)+\alpha^2+\alpha^2\cos(4\pi f_2 t)}{2} + 2\alpha\cos(2\pi f_1 t)\cos(2\pi f_2 t)\right\}$$

The last term is divided into $\alpha\cos(2\pi(f_1+f_2)t)+\alpha\cos(2\pi(f_1-f_2)\,t)$ of which amplitude is $\alpha\dfrac{I_{DSS}}{4} = \dfrac{2V_2}{|V_{GSoff}|}\dfrac{I_{DSS}}{4} = \dfrac{V_2}{|V_{GSoff}|}\dfrac{I_{DSS}}{2}$. The conversion slope (or conversion transconductance) is thus $\dfrac{I_{DSS}}{2|V_{GSoff}|} = 0.18$ mA/V or mS for the signal frequency $f_1 - f_2 = 100 - 89.3$ MHz = 10.7 MHz (intermediate frequency in the FM radio band).

3.5.2. *JFET analog switch; input–output insulation in off-state*

i) Show that when Q_2 is off, diode D_1 is on but with a voltage drop of zero (since the gate current of the JFET is negligible) and that Q_1 then remains off if V_1 meets a certain condition. Which?

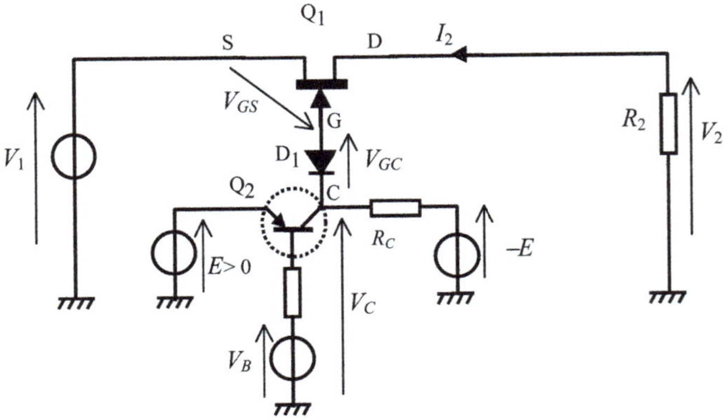

ii) Draw the equivalent dynamic circuit of the gate above while accounting for capacitance C_{GS}, C_{GD} (C_{DS} disregarded) of Q_1, of capacitance C_1 of diode D_1 and of R_C. Why the dynamic equivalent circuit of diode D_1 can be reduced to the capacitance C_1 alone?

iii) For simplicity, let us assume that $C_{GS} = C_{GD} = C_0 << C_1$ and $R_C = R_2 = R$, then calculate the ratio of variations $T(j\omega) = V_2/V_1$ in sinusoidal conditions at circular frequency ω. Show that $T(j\omega)$ takes the form

$$T(j\omega) = \frac{j\dfrac{\omega}{\omega_1}\left(1 + j\dfrac{\omega}{\omega_2}\right)}{\left(1 + j\dfrac{\omega}{\omega_3}\right)\left(1 + j\dfrac{\omega}{\omega_4}\right)}$$

Draw the asymptotic diagram of $\log|T(j\omega)|$ as a function of $\log \omega$. Numeric application: beyond what frequency does the asymptotic value of transmission $|T|$ go above – 40 dB for $C_0 = 4$ pF; $C_1 = 60$ pF and $R = 1$ kΩ?

Answer:

i) Diodes D_1 and G-channel are in series and in opposite connexion (back to back); thus they are traversed by zero current. The voltage drop of that which is not under reverse bias is thus zero. Since $V_C = -E$ (with $E > 0$), D_1 is then on if the value of V_1 remains above $-E$. If furthermore Q_1 is to remain off, we must impose $V_{GS} < V_{GSoff} = -V_p + \Phi < 0$. Since $V_{GS} = -E - V_1$, $V_1 > -E + V_p - \Phi = -E + |V_{GSoff}|$.

ii) Since diode D_1 is subject to zero voltage, its conductance can be disregarded as under reverse bias because the conduction threshold is not reached. So the equivalent dynamic circuit of diode D_1 can be reduced to capacitance C_1 alone. Since transistor Q_1 is off, its equivalent dynamic circuit is reduced to capacitances C_{GS} and C_{GD} alone. Since Q_2 is off, C_1 is returned to the dynamic ground through R_C. Hence, the equivalent dynamic schematics become:

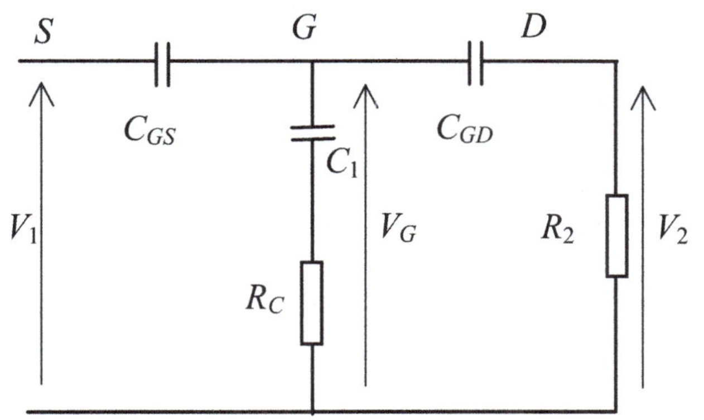

iii) We write out the equation at node G and the voltage divider giving V_2 to be able to remove V_G. Since $C_{GS} = C_{GD} = C_0 << C_1$, we have (Millman theorem):

$$V_G = \frac{jC_0\omega(V_1 + V_2)}{2jC_0\omega + \dfrac{jC_1\omega}{1 + jR_CC_1\omega}} \text{ and } V_2 = V_G\frac{jR_2C_0\omega}{1 + jR_2C_0\omega}$$

so: $$V_2 = \frac{jR_2C_0\omega}{1 + jR_2C_0\omega}\frac{jC_0\omega(V_1 + V_2)}{2jC_0\omega + \dfrac{jC_1\omega}{1 + jR_CC_1\omega}}$$

With $R_C = R_2 = R$, we see:

$$T(j\omega) = \frac{V_2}{V_1} = \frac{jRC_0\omega(1 + jRC_1\omega)}{2 + \frac{C_1}{C_0} + jR\omega(3C_1 + C_0) + (jR\omega)^2 C_1 C_0}$$

which can be simplified accounting for $C_0 << C_1$, thus yielding:

$$T(j\omega) = \frac{V_2}{V_1} = \frac{C_0}{C_1} \frac{jRC_0\omega(1 + jRC_1\omega)}{1 + 3jRC_0\omega + (jRC_0\omega)^2}$$

So we indeed have $T(j\omega) = \dfrac{j\dfrac{\omega}{\omega_1}\left(1 + j\dfrac{\omega}{\omega_2}\right)}{\left(1 + j\dfrac{\omega}{\omega_3}\right)\left(1 + j\dfrac{\omega}{\omega_4}\right)}$ with $\omega_1 = \dfrac{C_1}{RC_0^2}$,

$\omega_2 = \dfrac{1}{RC_1}$, hence $\omega_1 = \omega_2 \left(\dfrac{C_1}{C_0}\right)^2 >> \omega_2$ and

$$\omega_3 = \frac{1}{RC_0}\frac{3+\sqrt{5}}{2} = \frac{C_1}{C_0}\omega_2\frac{3+\sqrt{5}}{2}, \quad \omega_4 = \frac{1}{RC_0}\frac{3-\sqrt{5}}{2} = \frac{C_1}{C_0}\omega_2\frac{3-\sqrt{5}}{2}$$

So we have in increasing order $\omega_2 < \omega_4 < \omega_3 < \omega_1$, so that the following asymptotic diagram can be plotted in dB:

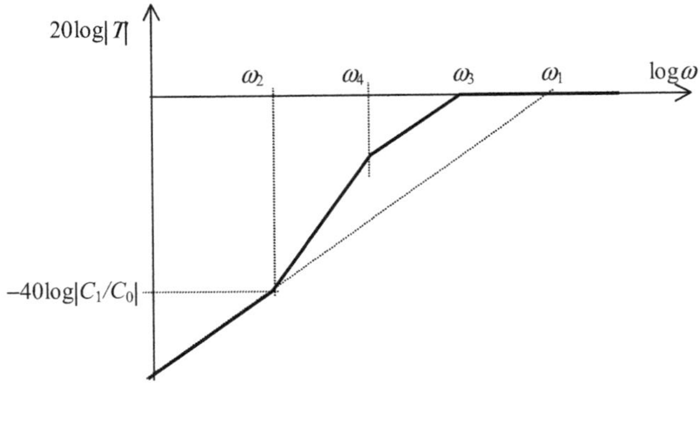

Numeric application with $C_0 = 4$ pF; $C_1 = 60$ pF and $R = 1$ kΩ:

$$f_2 = \frac{\omega_2}{2\pi} = 2.65 \text{ MHz}; \; f_4 = \frac{\omega_4}{2\pi} = 15.2 \text{ MHz}; \; f_3 = \frac{\omega_3}{2\pi} = 104 \text{ MHz}$$

and $f_1 = \frac{\omega_1}{2\pi} = 596 \text{ MHz}$. For $f = f_2$, the asymptotic value of transmission $|T|$ can be calculated by factor $\left| j\frac{\omega_2}{\omega_1} \right| = \left(\frac{C_0}{C_1} \right)^2$ alone since this still concerns the first asymptote, so $-40\log|C_1/C_0| = -47$ dB. The slope then becomes +40 dB/decade. So to reach –40 dB we have to multiply frequency f_2 by $10^{7/40}$ = 1.5, which provides a frequency of 4 MHz.

3.5.3. *MOSFET circuits*

i) Logic inverter

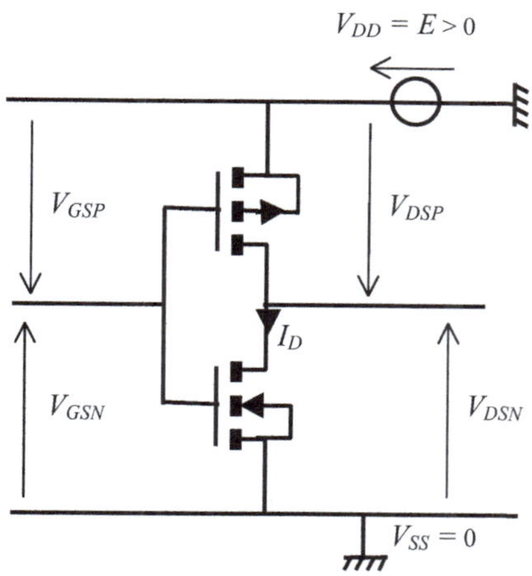

Assuming the characteristics of the NMOS and PMOS transistors to be perfectly symmetrical with respect to the origin, plot the network of characteristics $I_D(V_{DSN})$ that allows to determine the operating point in this inverter circuit. Then, draw the path of the operating point when V_{GSN} shifts from 0 to E.

Establish the dynamic small signal and low-frequency equivalent circuit, then determine the voltage gain around operating point $V_{DSN} = E/2$ as a function of the transconductance g_m and the output conductance $1/r_{DS}$. Numeric application: $g_m = 10^{-3}\ \Omega^{-1}$; $r_{DS} = 4 \times 10^5\ \Omega$.

ii) Establish the positive logic truth table of the circuit below:

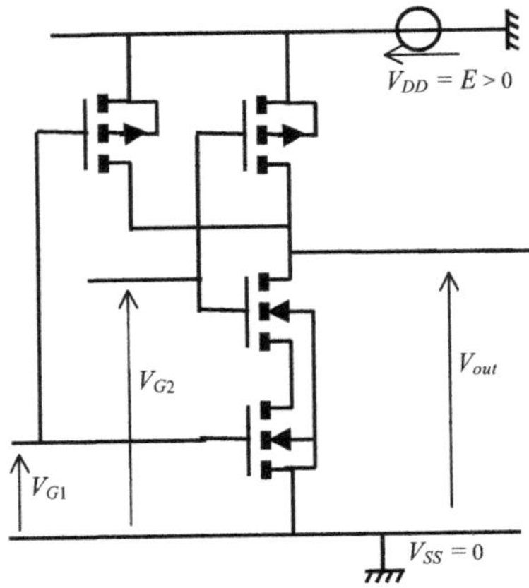

NOTE.– There is no connection at the wire crossings, only at the T contacts.

Answer:

i) Logic inverter:

With the usual assumption of currents entering the drain, we have the PMOS characteristic that is opposed to that of the NMOS since currents I_D and voltages V_{DS} have opposite signs.

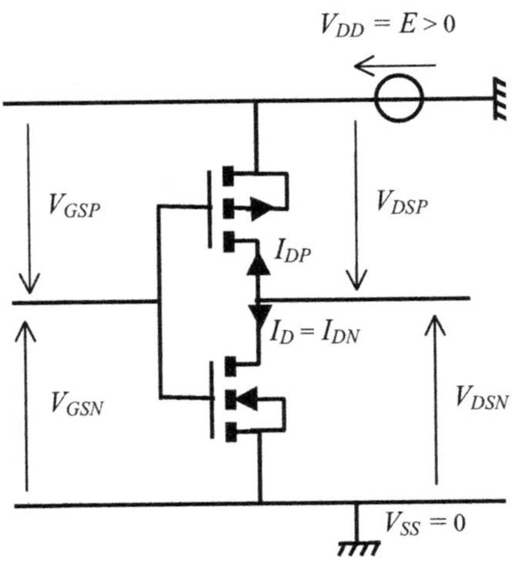

Hence characteristics are (with $V_{GS}1<V_{GS}2<V_{GS}3<V_{GS}4$ for the NMOS):

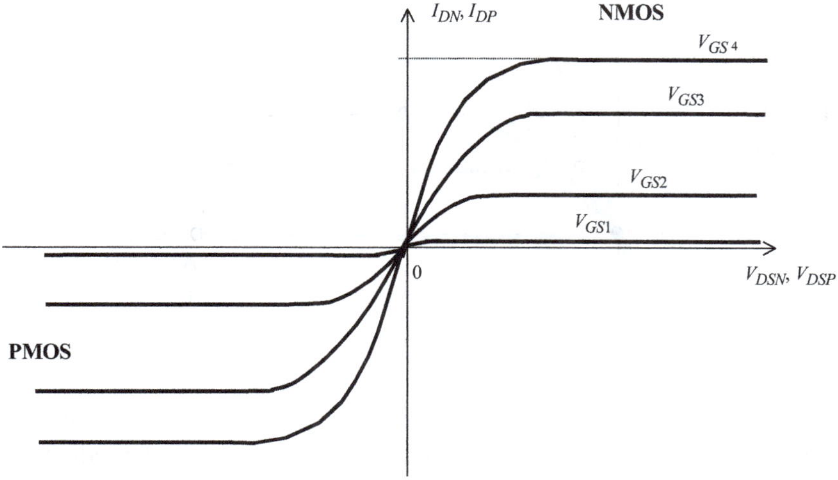

But since the drain current is common to both, we have $I_{\mathrm{D}} = I_{\mathrm{DN}} = -I_{\mathrm{DP}}$:

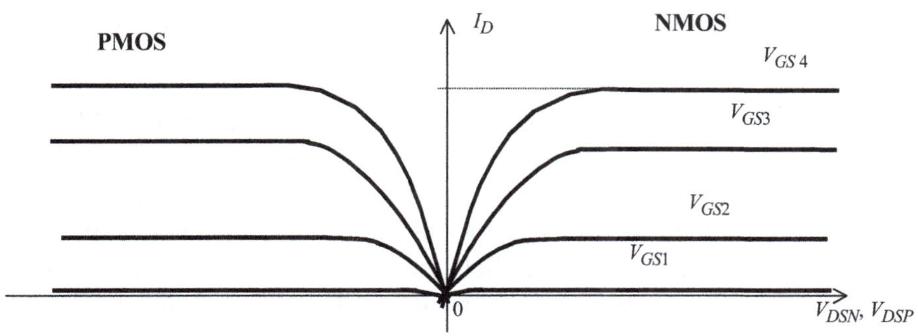

Finally, $E = V_{DSN} - V_{DSP}$, or an output voltage $V_{DS} = V_{DSN} = E + V_{DSP}$, which implies translating the PMOS characteristic of $+E$ to the right since $E > 0$. The operating point is at the intersection of the characteristics and follows voltage $V_G = V_{GSN} = E + V_{GSP}$ imposed on input.

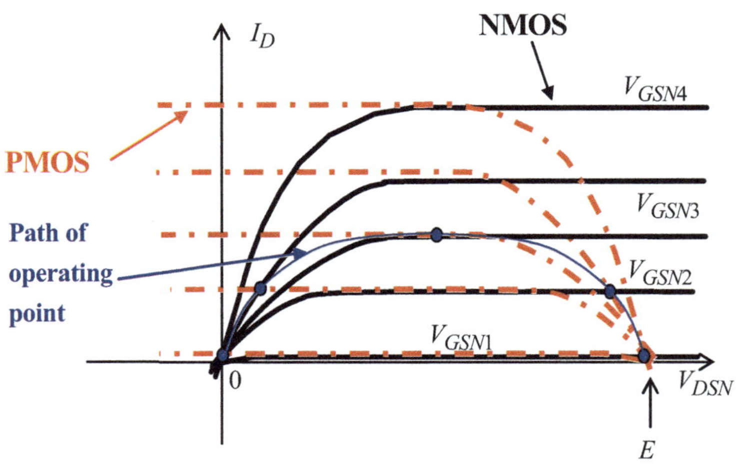

In dynamic conditions, we find both current sources $g_m\ v_{GS}$ (PMOS and NMOS) that converge to give a current 2 $g_m\ v_{GS}$ in two parallel conductances $1/\ r_{DS}$, which is an equivalent conductance $2/r_{DS}$. The voltage $-g_m\ r_{DS}\ v_{GS}$ results, yielding a voltage gain $-g_m\ r_{DS}$. It is this voltage gain that is used, for example, in a CMOS inverter oscillator using a quartz resonator (see Chapter 1 of Volume 2 [MUR 17a]).

ii) To establish the circuit's truth table, we must study the state of the transistors for the various combinations of V_{G1} and V_{G2}.

When V_{G1} or V_{G2} is at zero voltage (low level "0"), one of the NMOS is off or rather when both voltages are zero, both are off. Since they are in series, the current is necessarily zero in the branch including the NMOS. However, one of the two or both PMOS are always conductive in such conditions. Since they are in parallel, the voltage fall in the conductive transistor(s) is zero since there is no current, and $V_{out} = E$ (high level "1").

Conversely, if V_{G1} and V_{G2} are at high level ("1"), corresponding to voltage E, both NMOS are conductive and both PMOS off. So $V_{out} = 0$ (low level "0").

This operation is announced by a truth table expressed by the logical relation $V_{out} = \overline{(V_{G1}.V_{G2})}$, which is a NAND function.

iii) MOSFET switch

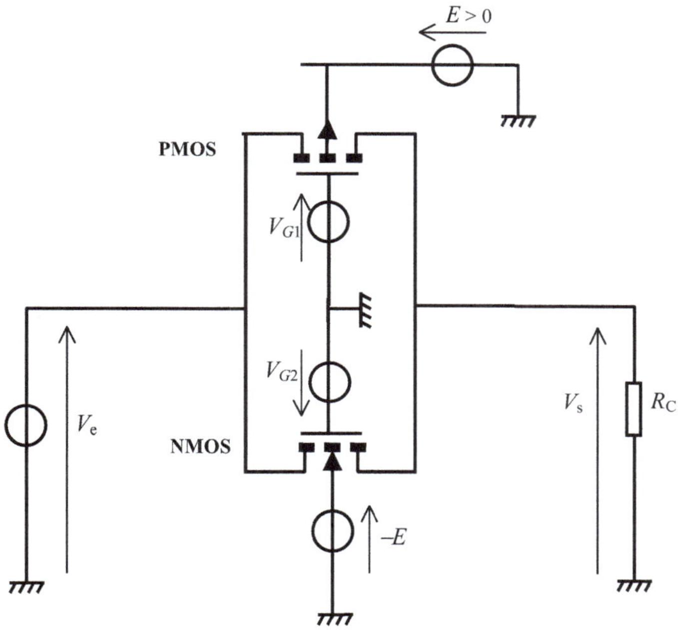

a) MOSFETs become conductive with holes in the channel for PMOS and electrons for NMOS, respectively, which implies charges of opposite signs on the gates. The gate-channel voltage will thus be negative for PMOS and positive for NMOS to both be conductive, so $V_{G1} = -E$ and $V_{G2} = +E$. In this case, we can as a first approximation consider that the source, channel and drain of a given transistor are at the same voltage, that is to say, neglecting the voltage drop in the conductive channel, and thus V_{GS} is also the gate-channel voltage difference. Note that in this case, the channel–substrate junctions are off as long as V_e does not stray from the $[-E, +E]$ range.

However, with gate voltages of opposite sign, both transistors are off, since for both there are diodes (source-channel and channel-drain) back to back while the substrate–channel diode disappears because the channel is no longer inverted (the PMOS channel becomes n-type as per its construction and inversely for NMOS because they are enhancement transistors). We can thus consider that the channel and substrate voltages are identical, which makes all the channel–source and channel–drain diodes off so long as V_e

does not stray from the $[-E,+E]$ range (since drains and sources are of opposite type to the initial type of the channel, which in order is source-channel-drain, that is pnp for the PMOS and npn for the NMOS).

b) Total conductance is the total of both ones and accounting for $V_{GS1} = V_{G1} - V_e$ and $V_{GS2} = V_{G2} - V_e$, we find that it is independent of V_e, which would clearly not be the case for a single MOSFET (we can also disregard V_{DS} as in (a)), with the transistors being conductive. To enter greater detail, in the conductive state, the variable resistance regime conductance is:

$$g_{DSN} = \frac{Z}{L}\mu_n C_{ox}\left(V_{GS2} - V_{DS} - V_{TN}\right), \text{ with } V_{TN} > 0 \text{ for NMOS}$$

and $g_{DSP} = \dfrac{Z}{L}\mu_p C_{ox}\left(V_{DS} + V_{TP} - V_{GS1}\right)$, with $V_{TP} < 0$ for PMOS.

Assuming $\mu_n = \mu_p = \mu$, $V_{TN} = -V_{TP}$, $V_{GS1} = V_{G1} - V_e$ and $V_{GS2} = V_{G2} - V_e$, we obtain:

$$g_{DSN} = \frac{Z}{L}\mu C_{ox}\left(V_{G2} - V_e - V_{DS} - V_{TN}\right), \text{ for NMOS}$$

and $g_{DSP} = \dfrac{Z}{L}\mu C_{ox}\left(V_{DS} - V_{TN} - V_{G1} + V_e\right)$, for PMOS.

The total of these conductances (since the transistors are parallel) is, accounting for $V_{G1} = -E$ and $V_{G2} = +E$:

$$g_{DSN} + g_{DSP} = 2\frac{Z}{L}\mu C_{ox}\left(E - V_{TN}\right)$$

In on-state, the conductance is thus constant, independent of voltage V_e and voltage drop V_{DS}, as this would occur in an ideal, passive component with constant resistance in on-state:

$$g_{DSN} + g_{DSP} = 2 \times 0.5 \times 10^{-3} = 10^{-3} \text{ S}$$

which is equivalent to a resistance of 1,000 Ω.

In order to decrease this resistance, we must simply increase the width Z of the gates (1 mm would give 1 Ω). This property would be absent with a single transistor.

The limitation on V_e in the off-state can be deduced from the conditions that maintain zero conductance or $V_{GS2} - V_{DS} - V_{TN} < 0$ and $V_{DS} + V_{TP} - V_{GS1} < 0$. Since in the off-state there is no voltage drop in R_C, we have $V_e = -V_{DS}$ and then:

For NMOS $(V_{G2} = -E)$: $-E + V_e - V_{TN} < 0$, which is $V_e < E + V_{TN} = 11$ V.

For PMOS $(V_{G1} = E)$: $- V_e + V_{TP} - E < 0$, which is $V_e > V_{TP} - E = -11$ V.

Finally, the switch remains in off-state so long as $-11 \text{ V} < V_e < 11 \text{ V}$ if the gate control voltages are of ±10 V together with the substrates' biasing voltages, and the transistors' threshold voltages of ±1 V.

4

Amplifiers, Comparators and Other Analog Circuits

4.1. Operational amplifiers, operating principle and types

4.1.1. *Standard operational amplifiers*

In linear regime, operational amplifiers provide an output voltage that is proportional to differential input voltage as long as both inputs remain within a voltage range somewhat narrower than the range of bias voltages V^+ and V^-. They generally involve three amplifier stages:

– a differential amplifier, often loaded by a current mirror;

– a high-voltage gain amplifier, generally with active load;

– a current amplifier with low output impedance, which can be considered in a first approximation as a voltage source, except when output current limitation is active (in case of output short circuit or too low a load resistance).

The set is completed by active bias circuits (current source and current mirrors). As an example, Figure 4.1 shows the schematics of one of the first generation operational amplifiers, built only with bipolar junction transistors.

Similar to all active components and resistors, operational amplifiers have sources of noise (voltages and currents of random instantaneous value), superimposed to the useful signal, as will be detailed in section 4.4. In order to reduce this noise and also the input currents, operational amplifiers with field effect transistors are now built (JFET in the first example in Figure 4.2; MOSFET in Figure 4.3).

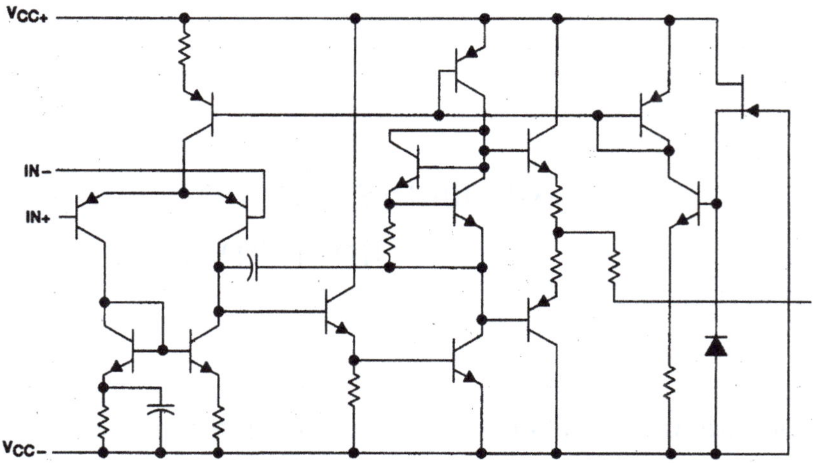

Figure 4.1. *Example no. 1, type 741*

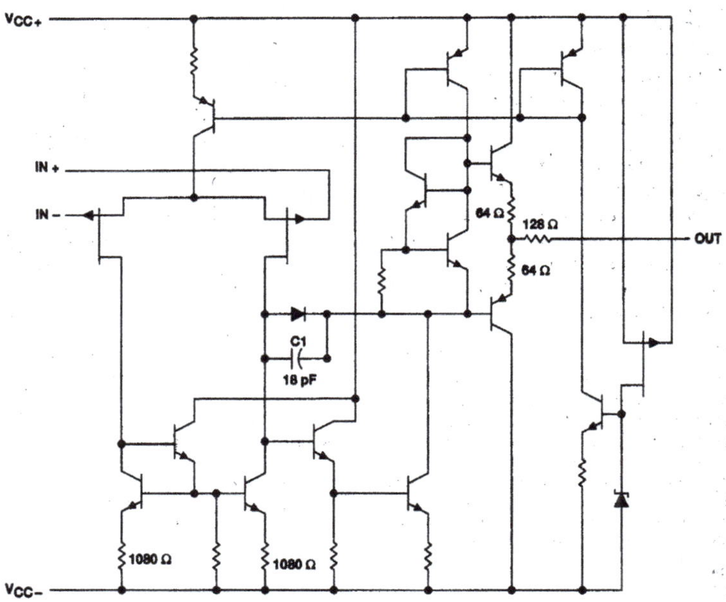

Figure 4.2. *Example no. 2, TL074 (STMicroelectronics, Harris, Philips, etc.) with p-channel JFET input*

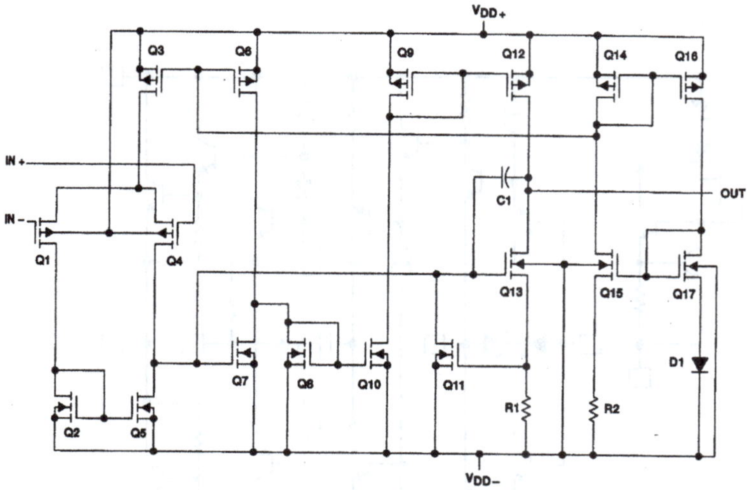

Figure 4.3. *Example no. 3, TLC2201 (Motorola) with p-channel and n-channel MOSFETs (CMOS)*

4.1.2. *Operational amplifiers with specific properties*

– "Rail to rail" operational amplifiers: Unlike the previous ones, two input voltages and/or the output voltage can reach the bias voltages V^+ and V^- (with mV precision) while preserving linear operation. They can then be readily used with $V^- = 0$ (and only one bias source, V^+).

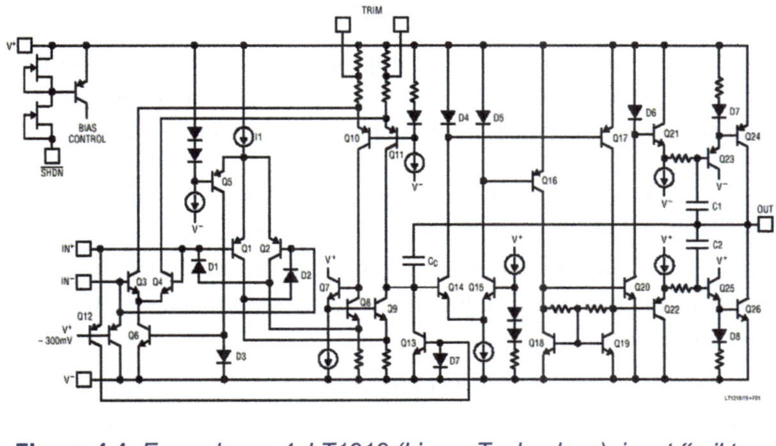

Figure 4.4. *Example no. 4, LT1218 (Linear Technology), input "rail to rail" operational amplifiers*

– Amplifiers operating with current feedback.

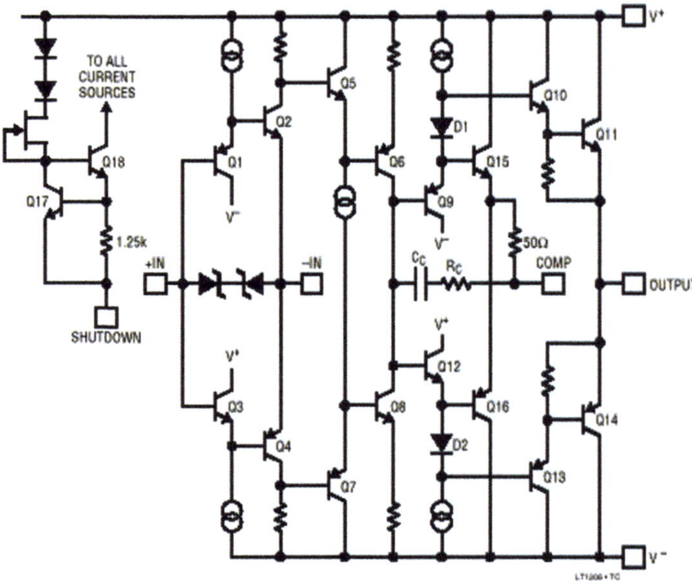

Figure 4.5. *Example no. 5, LT1206 (Linear Technology)*

These are rapid amplifiers with gain-bandwidth product that can easily exceed the gigahertz. The input resistance on minus input (–IN) is very low (generally below 20 Ω), as these are Q2-Q4 emitters, equivalent to common base inputs, while it reaches several kiloohms on plus input (+IN), for Q1-Q3 bases, as shown by the example in Figure 4.5. These are also called transresistance (or transimpedance) amplifiers, as the output voltage is proportional to the current in the minus input, the proportionality factor being equivalent to a resistance, namely the transresistance.

– Transconductance or transadmittance amplifiers: Input stages have quite high input impedance (>100 kΩ) but the output delivers a current that is proportional to the differential input voltage in linear regime, the proportionality factor being equivalent to a conductance, namely the transconductance.

– Differential output amplifier (Figure 4.6): These amplifiers are not to be found in individual packages, but rather integrated in switched capacitor circuits, as they serve a very specific interest. Indeed, in CMOS technology, in switched capacitor circuits, it is useful to eliminate offsets due to parasitic capacitances, which can be achieved by symmetrizing the feedback loops of operational amplifier assemblies. This can only be done with a differential output.

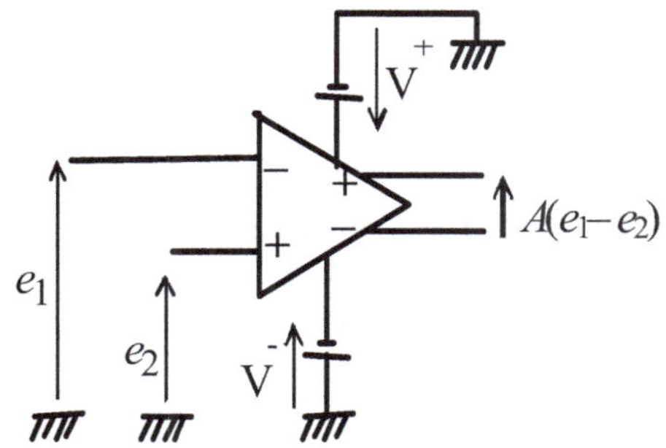

Figure 4.6. *Differential output operational amplifier*

4.2. Operational amplifier models and responses

4.2.1. *Static model of voltage amplifier*

The following characteristic curve relating output voltage V_s to input differential voltage ε (equal to the difference of input voltages e^+ and e^-) is obtained for a standard operational amplifier in quasi-static operation.

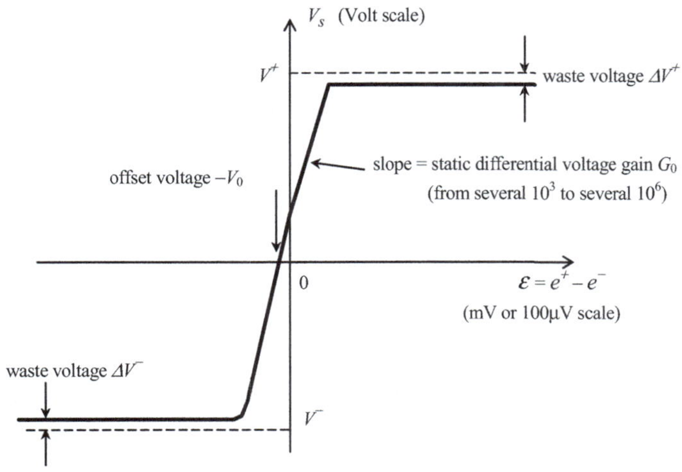

Figure 4.7. *Static transfer characteristic curve of operational amplifier*

Static imperfections are essentially equivalent input offsets:

– static voltage V_0 input offset;

– V_{EMC}, input offset due to non-zero common mode gain that adds a common mode offset voltage equal to the common mode rejection ratio (CMRR) multiplied by the common mode voltage $(e^+ + e^-)/2$. CMRR is defined as the ratio of common mode gain G_{MC} and differential gain G_0 or most often by its equivalent in dB;

– offset due to input currents flowing through external resistances, specified by the data of average input bias current $I_B = \dfrac{I_E^+ + I_E^-}{2}$ and differential current expressed in absolute value $I_D = \left|I_E^+ - I_E^-\right|$.

The resulting static or very low frequency equivalent circuit is shown in Figure 4.8.

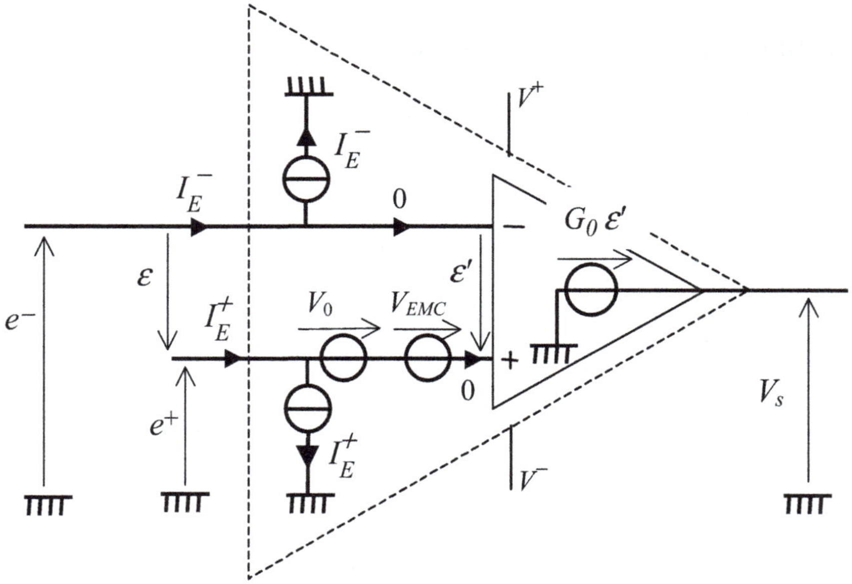

Figure 4.8. *Equivalent static and low-frequency circuit of operational amplifier*

Static voltage gain G_0 in open-loop ranges from about 10^4 to around 10^7 depending on the amplifier reference.

4.2.2. *Dynamic and switched mode operations*

The complete equivalent circuit in dynamic mode will take into account the differential input and common mode impedances, the output resistance, the differential gain $G_0(f)$, the common-mode rejection ratio CMRR(f) and the rejection ratios fed back to input $G^+(f)$ and $G^-(f)$ for supply voltage variations ΔV^+ and ΔV^-, all depending on the frequency.

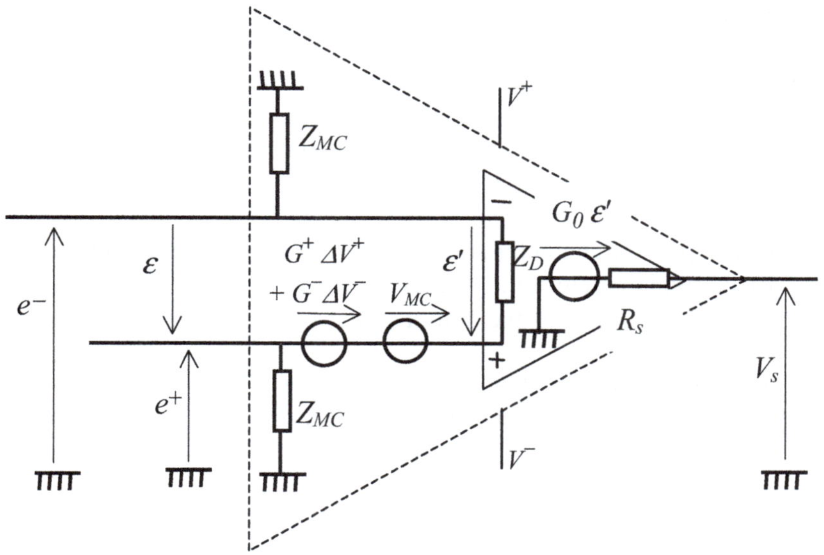

Figure 4.9. *Equivalent dynamic circuit taking into account differential and common mode imperfections as well as perturbations due to bias voltage fluctuations*

Highlighting supply voltage and common-mode rejection ratios, output voltage is given by:

$$V_s = G_0\left(\varepsilon + \frac{G^+}{G_0}\Delta V^+ + \frac{G^-}{G_0}\Delta V^- + \frac{G_{MC}}{G_0}V_{MC}\right)$$

The last three terms in the right-hand side of the equation are perturbations affecting the ideal response, which depends only on differential voltage ε and corresponds exclusively to the first term.

Frequency response is mainly determined by that of $G_0(f)$, of low-pass type, generally of first order, often characterized by unity gain frequency, equal to $G_0(0)$ multiplied by the cut-off frequency f_c. The output voltage V_s variation rate is limited by the slew rate, upper bound of $\left|\frac{dV_s}{dt}\right|$, which influences the response to strong signals and in switched mode operation.

4.3. Comparators

Comparators are operational amplifiers specifically optimized to perform the sign detector function. They allow the comparison between two voltages and the output delivers a logical level H or L that reflects the comparison outcome. For example: $\varepsilon > 0 \Rightarrow V_s = \text{H}$; $\varepsilon < 0 \Rightarrow V_s = \text{L}$.

The most important characteristics are the differential static gain and the voltage offset, which determine the accuracy, and the slew rate or the switching time that conditions the output signal variation rate.

Possible types of output stage include TTL or CMOS stages and open collector (and possibly emitter) transistors, as shown in Figures 4.10 and 4.11.

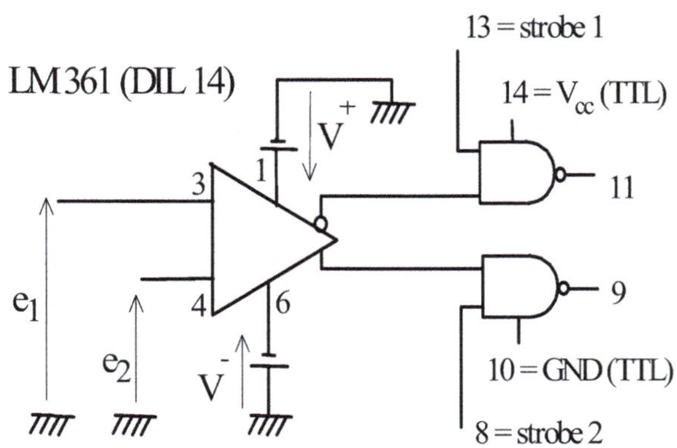

Figure 4.10. *Comparator with complementary TTL output stage and inhibition control*

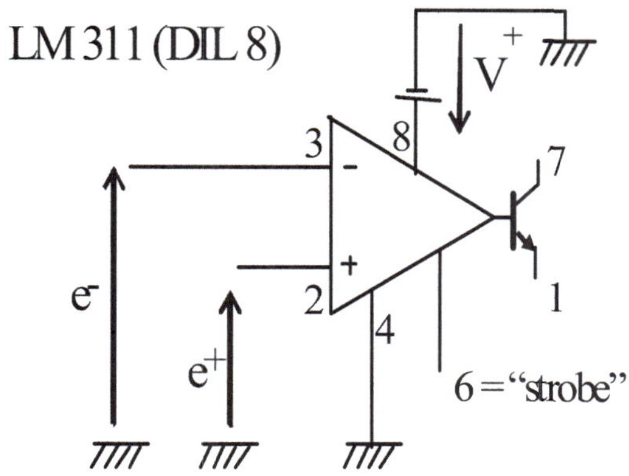

Figure 4.11. *Comparator with open collector and emitter outputs*

4.4. Noise in amplifiers

4.4.1. *Noise nature and evaluation*

Noise is a random variation of current or voltage that interferes with bias quantities and useful signal. Its instantaneous value is not predictable and must be considered a random variable whose possible values have definable probability. But a more resourceful approach to quantitatively defining noise involves the introduction of its autocorrelation function, which measures the similarity between instantaneous values separated by variable time lag. Having introduced a time scale, this autocorrelation function allows for the definition of an effective value of current noise and voltage noise, which is frequency dependent in the most general case. More specifically, this involves the use of noise power apart from a resistance factor and the square root of its derivative with respect to frequency, called noise spectral density, whose relations with the root mean squared (rms) values of noise voltage e_n and noise current i_n (n is the abbreviation for noise) are the following:

$$\eta_v = \sqrt{\frac{de_n^2}{df}}; \qquad \eta_i = \sqrt{\frac{di_n^2}{df}}$$

The most common units for these spectral densities are nV/$\sqrt{\text{Hz}}$ and pA/$\sqrt{\text{Hz}}$. When these spectral densities are constant and do not depend on frequency, the noise is called "white noise". On the other hand, spectral densities that vary with frequency are characteristic to "colored" noise.

Effective rms values of voltage and current e_b and i_b can be calculated within a frequency band $\Delta f = f_2 - f_1$ by:

$$e_b = \sqrt{\int_{f_1}^{f_2} \eta_v^2 \, df} \ ; \quad i_b = \sqrt{\int_{f_1}^{f_2} \eta_i^2 \, df}$$

For an amplifier, these values depend on bandwidth.

When several sources of voltage noise in series e_{b1}, e_{b2} ... and current noise in parallel i_{b1}, i_{b2} ... are going through a resistance R, the additivity rule will be applicable to rms values, yielding a total noise:

$$e_{btot} = \sqrt{e_{b1}^2 + e_{b2}^2 + \cdots + Ri_{b1}^2 + Ri_{b2}^2 + \cdots}$$

4.4.2. *Various types of noise and their origin*

4.4.2.1. *Thermal noise in a resistor*

Due to thermal agitation of charged particles carrying electric current, a resistor generates thermal noise or Johnson noise, even in the absence of current. This is a white noise up to frequencies of the order of hundreds of THz (infrared frequencies), whose spectral density is $\sqrt{4kTR}$ if the equivalent source of voltage noise is in series with the resistance R (conductance $G = 1 / R$) or $\sqrt{4kTG}$ if the equivalent source of current noise is in parallel with it, with $k = 1.38 \times 10^{-23}$ J/K, the Boltzmann constant, and T the absolute temperature in degrees Kelvin. For $R = 1$ kΩ and $T = 300$ K, these values are 4nV/$\sqrt{Hz}$ and 4nA/$\sqrt{Hz}$, respectively.

Therefore, rms noise voltage is $\sqrt{4kT\,R\,\Delta f}$ and rms noise current is $\sqrt{4kT\,G\,\Delta f}$.

Since there are no resistors in integrated circuits, this type of noise should not exist. Nevertheless, linearized models of elementary components feature resistors, and this is why thermal noise is generated inside each component. It is also considered as reference, particularly because the external source supplying input signal is never ideal and has internal resistance or conductance.

4.4.2.2. *Shot noise*

This noise is related to the particle nature of current, which is formed of charged particles (of elementary charge q) that are subjected to collisions and do not all simultaneously cross a given area of the conductor. For a current I and within a time interval Δt, the average number of charge carriers crossing this area is $I\,\Delta t\,/q$, with a Poisson probability whose variance is equal to this average number. In order to calculate the variance of current I, this average number should be multiplied by coefficient $q/\Delta t$ squared, giving $Iq/\Delta t$, which yields a power spectral density equal to Iq. It is constant and therefore corresponds to white noise. If spectral foldover is taken into account, the mean squared noise current is $2qI\Delta f$ and the spectral density is $\eta_{I0} = \sqrt{2\,q\,I}$.

This noise is present in all components and its spectral density can be modified by carrier trapping and diffusion, resulting in deviation from the previous value. Beyond a certain corner frequency f_{c2}, spectral density of noise current generally increases in $\eta_I = \eta_{I0}\sqrt{1+\left(\frac{f}{f_{c2}}\right)^2}$ with frequency f, but the exponent can slightly deviate from the standard value.

4.4.2.3. *"1/f noise" and other noises*

$1/f$ noise is defined by the frequency dependence of its spectral density rather than by its physical cause, the latter being unknown when this noise was discovered. It was subsequently acknowledged that this noise emerged at the interfaces present in components, particularly in field effect transistors, in which conduction takes place in the proximity of the interface between two types of the same semiconductor (JFET) or between two different semiconductors (heterojunction FET), between a metal and a semiconductor (MESFET) or between an insulator and a semiconductor

(MOSFET). In this latter case, voltage fluctuations and carrier trapping and untrapping in and out of the interface states are responsible for the highest spectral density. This is the reason why JFET and MESFET are generally preferred in low-noise amplifier (LNA) design. Spectral density of noise voltage is in fact constant beyond a certain frequency f_{c1} of the order of several tens of hertz ($f_{c1} < f_{c2}$), lower for JFET than for MOSFET, all things being equal, and can be written as $\eta_{1/f} = \eta_c \sqrt{1+\left(\frac{f_{c1}}{f}\right)^2}$, while the exponent may slightly deviate from the standard value. This noise is named or confused with flicker noise. Other noises of similar origin are known, such as the telegraph noise or "popcorn" noise, with sudden step-like variations of offset voltage or current.

All of these noises being linked to the particle nature of charge carrier and to the statistical nature of electric current, spectral densities increase when component size diminishes, as the relative value of fluctuations increases.

4.4.3. *Equivalent circuit of noise sources in amplifiers and noise figure*

Noises due to semiconductor components in an integrated circuit can be considered fluctuations of bias currents and offset voltage. The naturally resulting equivalent circuit is then the same as the one used for taking into account these static quantities (Figure 4.12).

This principle is also applied when a single transistor is used as amplifier, and in this case only one source of noise voltage and one source of noise current are added to the dynamic transistor circuit.

The characteristics provided by manufacturers indicate the spectral densities for these so-called "input equivalent" noise sources. Taking into account the equivalent resistances on each input, their overall effect can be calculated by the rule valid for rms values (see section 4.4.1), which allows for comparison with useful input signals. This operation involves the noise figure, which compares the noise power due to internal resistance of the input signal source to the total noise power at amplifier input:

Noise figure F = (noise power of the component + source)/(source noise power). Since:

$$F = \frac{\text{(Power of input signal)/(source noise power)}}{\text{G} \times \text{power of input signal/(G} \times \text{whole noise power of the component + source)}}$$

where G stands for power gain, the noise figure also measures the signal-to-noise ratio degradation between output and input, as the numerator corresponds to signal/noise ratio at input and the denominator to signal/noise ratio at output. Most of the time, it is given in dB according to $F_{dB} = 10 \log F$.

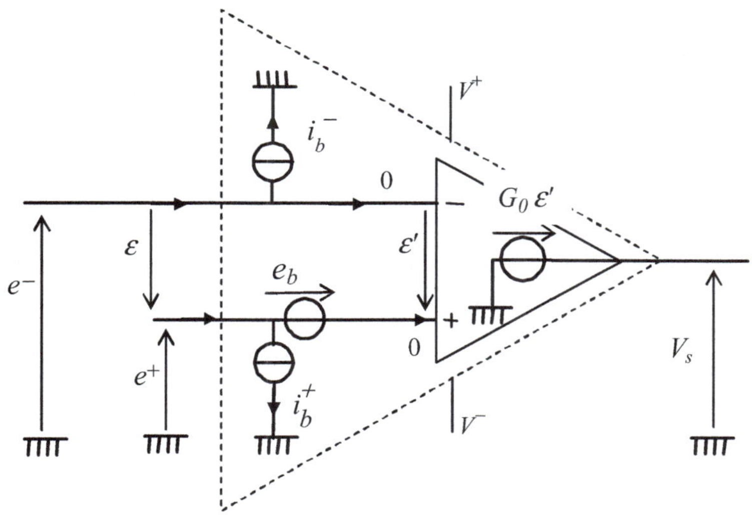

Figure 4.12. *Sources of noise fed back to operational amplifier input*

If e_b and i_b are, respectively, input noise voltage and current, it can be shown that F is minimum for an internal source resistance $R_{source} = e_b/i_b$ (see exercise), which is the case of noise adaptation.

These notions are all the more important as the signals to be amplified are weak, as in instrumentation and broadcasting receivers. When there are several amplification stages (numbered 1, 2, 3,…), it can be shown that the total noise figure is given by $F_1 + \frac{F_2 - 1}{G_1} + \frac{F_3 - 1}{G_1 G_2} + \cdots$ (Friis formula, in which noise figures and power gains are expressed in numerical values, and not in dB). It is therefore the noise figure of the first stage that has a dominant impact on the total noise figure, therefore it is important to minimize it by means of a low-noise preamplifier.

4.4.4. *Low-noise amplifiers*

Another case of weak signals, often well below μV, is encountered when the source is an antenna. These radiofrequency signals carry information in all telecommunication systems and have to be amplified in one or more stages at receiver input. Before 1990, LNAs often had two transistor stages. Due to subsequent progresses with respect to transistor performance in terms of bandwidth and noise figure, it is now possible to have one stage amplifiers, as common emitter (bipolar transistor) or common source (FET) in general. These transistors are either AlGaAs/GaAs heterojunction bipolar transistors (HBT) or more recently SiGe/Si HBT, or GaAs field effect transistors of MESFET, PHEMT (pseudomorphic high mobility transistor) or more recently submicron Si MOSFET type. The load is sometimes active (cascode) or a coupling of two transistors by emitters or sources can be used. Negative feedback is also possible. In all cases, design and modeling take into account both the circuit, which contains the passive components and, which serves as inseparable support for the active component, and the active component itself. Dimensions are all the smaller as the operating frequency is high. The notions specific to radiofrequency amplifiers will not be developed here. Operating frequencies range from several hundred megahertz to around 50 GHz depending on application (mobile telephone, GPS, telecommunication network node, etc.). A review conducted by D. K. Shaeffer and T. H. Lee can be found in [SHA 97] and further information on characteristics of radiofrequency components is available in the datasheets provided by specialized manufacturers.

4.5. Analog integrated circuits

While not exhaustive, Table 4.1 provides a list of possible functions of analog integrated circuits, in addition to those already described, accompanied by several details on the basic functions used in each case. This table does not include circuits related to the following chapters, which are dedicated to electronic systems.

Name of specialized circuit	Functions performed	Basic functions involved
Amplifiers	– Operational – Power – Instrumentation – Insulation	Transistor amplification stages, possibly optocoupler or transformer in the latter case
Voltage regulators, supply control and voltage detectors, CC-CC converters, voltage inverters, etc.	Functions allowing continuous supply of all circuits	Amplifier, reference source, analog switchers, etc.
Voltage–frequency and frequency–voltage converter	–	Single-shot flip-flop, integrator, comparator, etc.
Relaxation oscillator, time switches, function generators	Generator of various signals	Comparator, integrator, shaping circuitry, multiplier, etc.
Optocouplers	Insulated transmission of binary data	Light-emitting diodes and light-sensitive cell or phototransistor
Converter for signal shaping, drivers and line-receiver units	Analog transmission of signals emitted by sensors or interfaces	Amplification with output or input impedance matching
Data transmitters on power network	Encoded and decoded data transmission, generally in digital form	Carrier signal generation, encoding, modulation or demodulation
Infrared transmitters		Encoding, modulation or demodulation, LED driver
RF emitter, RF receiver		RF amplification, frequency change, intermediate selective amplification, modulation or demodulation

Table 4.1. *Some examples of functions achieved by specialized analog integrated circuits*

4.6. Exercises

4.6.1. *Responses of operational amplifier differential stage*

In the next circuit, sources E, $-E$ and I_{B0} are ideal direct bias sources (they supply voltages E and $-E$ that are <u>*constant*</u> or current I_{B0} that is *constant*). Bipolar junction transistors operate in conductive state and are identical. Field effect transistors operate in pinch-off state and are identical.

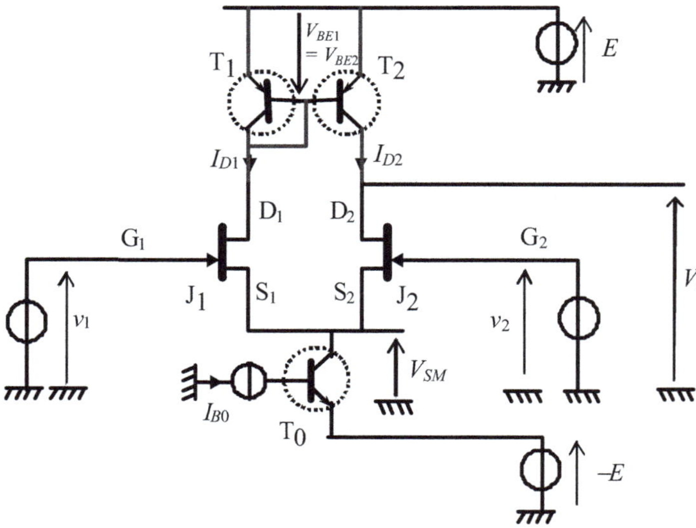

i) The linearized dynamic equivalent circuit (valid for small *variations* of current and voltage under low frequency near the operating point) for bipolar transistors is supposed to comprise a dynamic resistance r_{BE} crossed by i_B and an ideal current controlled source of current βi_B in parallel to a conductance g_{CE} (parameters r_{BE}, β and g_{CE} are identical for T_0, T_1, and T_2) connected to each collector; for field effect transistors, the equivalent circuit is composed only of an ideal voltage controlled source of current $g_m v_{GS}$ connected to each drain, where g_m is transistor transconductance (identical for J_1 and J_2) and v_{GS} is the gate voltage variation of the respective source (voltage v_{G1S1} or v_{G2S2}). Establish the equivalent dynamic circuit by representing the *variations* of current i_{B1}, i_{B2}, βi_{B1}, βi_{B2}, those in field effect transistors as a function of g_m and variations of voltage v_1, v_2, v_{SM} and v_D that represent variations of V_{SM} and V_D, respectively.

ii) Show that $i_{B1} = i_{B2} = i_B$. Then, write the three node equations for this dynamic circuit, for D_1 (equation (1)), S_1S_2 (equation (2)) and D_2 (equation (3)), as a function of variables i_B, v_1, v_2, v_{SM}, v_D and transistor parameters β, g_{CE} and g_m. In equation (1), the current through conductance g_{CE} of T_1 will be considered negligible, supposing g_{CE} r_{BE} <<1, which also entails a null variation of voltage at D_1.

iii) In common mode, $v_1 = v_2$. Find in this case the literal expression for common mode gain $G_{MC} = \dfrac{v_D}{v_1}$ as a function of transistor parameters, using the difference (1)–(3) to obtain i_B as a function of v_D, then (2) to obtain v_{SM} as a function of v_1, and finally (1) or (3). Calculate this gain when $\beta = 200$, $g_{CE} = 10^{-6}\ \Omega^{-1}$ and $g_m = 10^{-3}\ \Omega^{-1}$.

iv) In differential mode, $v_1 = -v_2$. Find in this case the literal expression for differential gain $G_D = \dfrac{v_D}{2v_1}$ as a function of transistor parameters, after having shown that $v_{SM} = 0$ according to (2) and eliminate i_B between (1) and (3). Calculate this gain for the same numerical values previously indicated. Deduce CMRR = G_D/G_{MC} and the corresponding value in decibels.

v) To find the differential gain frequency response (with $v_1 = -v_2$), capacitances C_{GD} and C_{GS} of J_1 and J_2, and $C_{B'E}$ and $C_{B'C}$ for T_1 and T_2 are taken into account, and D_1 potential is considered equivalent to the one of a dynamic ground because of the weak impedance of circuits BE of T_1 and T_2. The following dynamic circuit results:

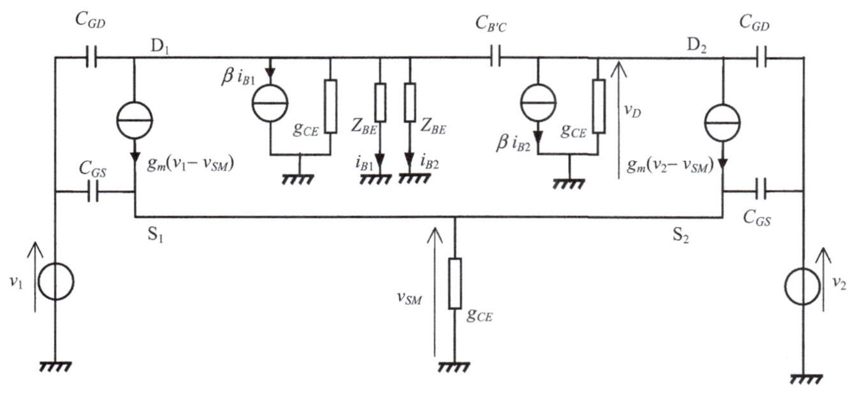

Write once again three node equations and show that $v_{SM} = 0$, similarly to the previous question. Considering $\beta+2 \approx \beta$, find the literal expression of G_D as a function of g_{CE}, g_m, C_{GD}, $C_{B'C}$ and $j\omega$, and calculate the time constants if $C_{GD} = C_{B'C} = 5$ pF.

Deduce the cut-off frequency for this differential gain. On the other hand, if $C_{B'E} = 100$ pF and $r_{B'E} = 5$ kΩ, justify that the existence of capacitances $C_{B'E}$ has been neglected in the above dynamic circuit.

Answer:

i) Dynamic equivalent circuit:

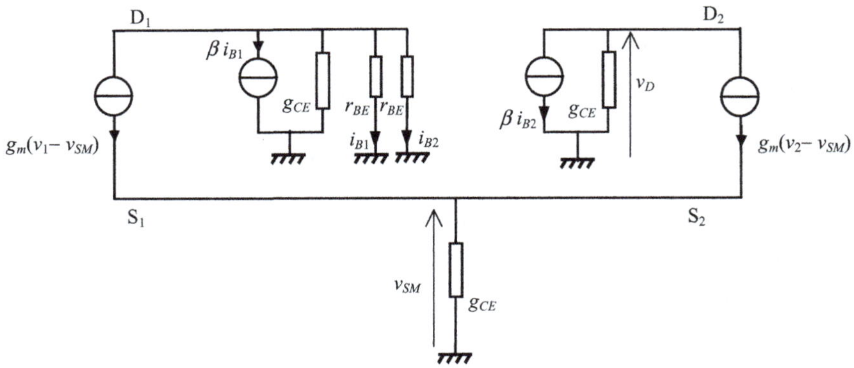

ii) $i_{B1} = i_{B2} = i_B$ since the same voltage applies across the two identical resistances r_{BE}. The three node equations of this dynamic circuit in D_1 (equation (1)), S_1S_2 (equation (2)) and D_2 (equation (3)) are:

$$g_m(v_1 - v_{SM}) + (\beta + 2)\, i_B + g_{CE}\, r_{BE}\, i_B = 0 \qquad (1)$$

$$g_m(v_1 - v_{SM}) + g_m(v_2 - v_{SM}) = g_{CE}\, v_{SM} \qquad (2)$$

$$g_m(v_2 - v_{SM}) + \beta\, i_B + g_{CE}\, v_D = 0 \qquad (3)$$

The last term of (1) is considered negligible, since $g_{CE}\, r_{BE} << 1$.

iii) In common mode, $v_1 = v_2$ and (1)–(3) then lead to $2\, i_B = g_{CE}\, v_D$. Relations (2) and (3) lead to $v_{SM} = \dfrac{2g_m}{g_{CE} + 2g_m} v_1$ and, after eliminating v_{SM}, the expression of common mode gain is obtained from $G_{MC} = \dfrac{v_D}{v_1}$:

$$G_{MC} = \frac{v_D}{v_1} = -\frac{2g_m}{(\beta + 2)(g_{CE} + 2g_m)}.$$

This gain is 5×10^{-3} in absolute value when $\beta = 200$, $g_{CE} = 10^{-6}\ \Omega^{-1}$ and $g_m = 10^{-3}\ \Omega^{-1}$.

iv) In differential mode, since $v_1 = -v_2$, the only remaining terms in equation (2) are those proportional to v_{SM}, therefore: $v_{SM} = 0$. Then, $(\beta+2)\, i_B + g_m v_1 = 0$ results from (1), which leads in (3) to: $2(\beta+1)\, i_B + g_{CE} v_D = 0$. This yields: $G_D = \dfrac{v_D}{2v_1} = \dfrac{(\beta+1)\, g_m}{(\beta+2)\, g_{CE}} = 10^3$ approximately.

The absolute value of CMRR is $CMRR = \dfrac{G_D}{G_{MC}} = \dfrac{(\beta+1)(g_{CE}+2g_m)}{2\, g_{CE}}$ which is 2×10^5 or 106 dB, which can be found as a typical value in the datasheet of operational amplifiers.

v) Given the following dynamic circuit:

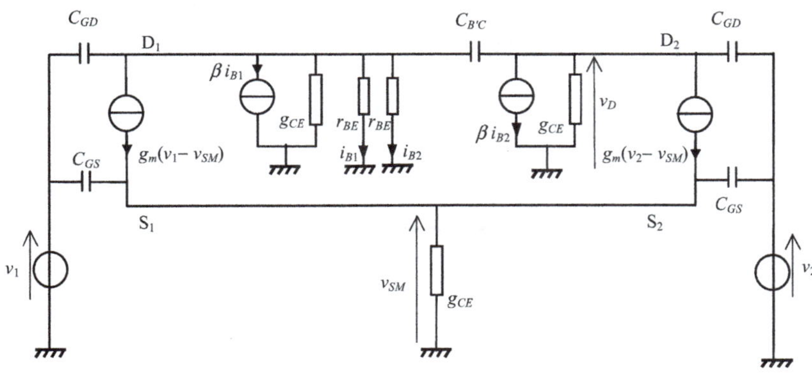

the three node equations are modified as follows:

$$g_m(v_1 - v_{SM}) + (\beta+2)\, i_B - j\, C_{B'C}\, \omega v_D - j\, C_{GD}\, \omega v_1 = 0 \tag{1}$$

$$(g_m + j\, C_{GS}\, \omega)(v_1 - v_{SM}) + (g_m + j\, C_{GS}\, \omega)(v_2 - v_{SM}) = g_{CE}\, v_{SM} \tag{2}$$

$$g_m(v_2 - v_{SM}) + \beta\, i_B + (g_{CE} + j\, C_{B'C}\, \omega)\, v_D + j\, C_{GD}\, \omega (v_{SM} - v_2) = 0 \tag{3}$$

Equation (2) yields v_{SM} when $v_1 = -v_2$, which allows the simplification of (1) and (3). Then, i_B is eliminated and considering $\beta + 2 \approx \beta$, we obtain $G_D = \dfrac{v_D}{2v_1} = \dfrac{g_m - jC_{GD}\,\omega}{g_{CE} + j\omega(C_{GD} + 2C_{B'C})}$. The value of time constants is 5 ns at the numerator and 15 μs at the denominator if $C_{GD} = C_{B'C} = 5$ pF. It is the one at denominator that determines the cut-off frequency of differential gain, which is around 10.6 kHz.

On the other hand, if $C_{B'E} = 100$ pF and $r_{B'E} = 5$ kΩ, the time constant corresponding to circuits BE of two transistors T_1 and T_2 is 1 μs << 15 μs, which justifies the approximation in the dynamic circuit.

4.6.2. *Generation of triangle and square wave signals: voltage to frequency conversion*

Study the $V_2(V_1)$ characteristic curve for the comparator circuit sketched below, in which $R_2 > R_1$, supposing that the operational amplifier transfer characteristic is the following and writing ε as a function of V_2 and V_1 (input current is considered negligible).

Then, analyze the evolution of ε and V_2 for values of V_1 ranging from V^- to V^+, then from V^+ to V^-. Draw the cycle $V_2(V_1)$ indicating the flow directions.

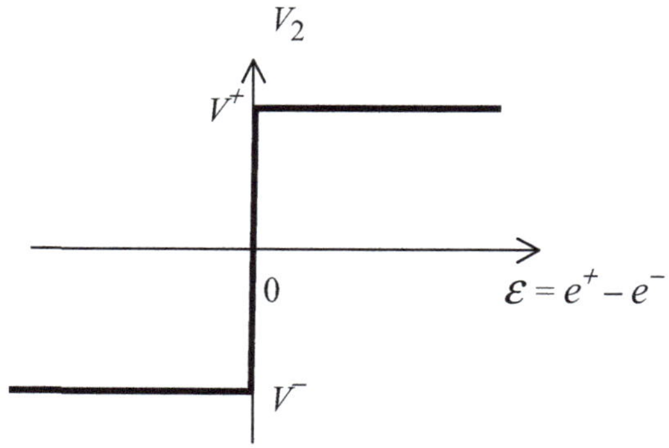

The source of voltage V_1 is now a voltage controlled source of voltage $V_1 = V_C$, where V_C is the voltage across capacitor C supplied by sources of current I_0 in the following schematics.

Inverter K (implemented with field effect transistors) is controlled by the output state $V_2 = V^-$ or $=V^+$.

Find the relation between V_C and I_0 depending on the state, either 1 or 2, of inverter K.

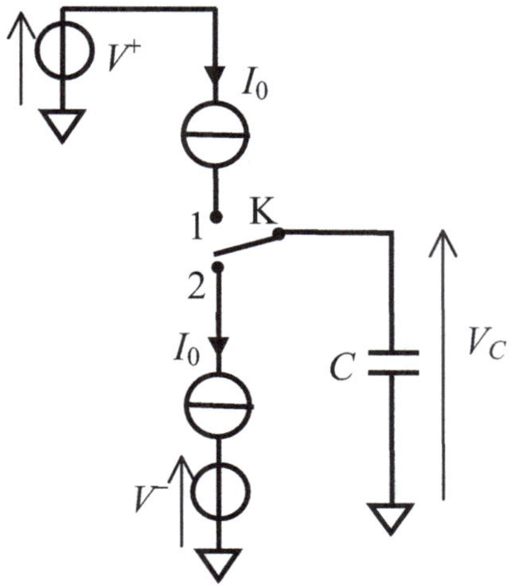

Then find the relation between states 1 or 2 of inverter K and the value of $V_2 = V^-$ or V^+ for the system to oscillate. Draw V_2 and V_C in synchronism.

Calculate the oscillation frequency f_2 as a function of the elements in the two circuits.

Show that if the source of current I_0 becomes a current source controlled by a voltage V_e (that is $I_0 = V_e / R_0$), the result is a voltage–frequency converter in which f_2 is proportional to V_e. Find the expression of the coefficient of proportionality.

Answer:

i) For the comparator circuit shown below:

$$\varepsilon = \frac{R_2}{R_1 + R_2} V_1 + \frac{R_1}{R_1 + R_2} V_2$$

If $\varepsilon > 0$, $V_2 = V^+$, therefore it implies $V_1 > -\frac{R_1}{R_2} V^+$ (a).

If $\varepsilon < 0$, $V_2 = V^-$, therefore it implies $V_1 < -\frac{R_1}{R_2} V^-$ (b). Moreover, $\frac{R_1}{R_2} < 1$.

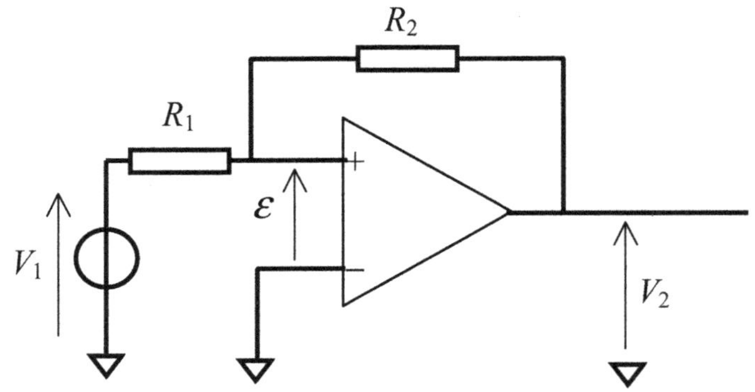

Therefore if V_1 starts from $V^- < 0$, condition (b) is fulfilled. If V_1 exceeds the positive value $U^+ = -\frac{R_1}{R_2}V^-$, ε becomes positive inducing $V_2 = V^+$. Condition (a) is then fulfilled and is unchanged as long as V_1 does not fall below $U^- = -\frac{R_1}{R_2}V^+ < 0$.

Hysteresis comparator operation is then obtained.

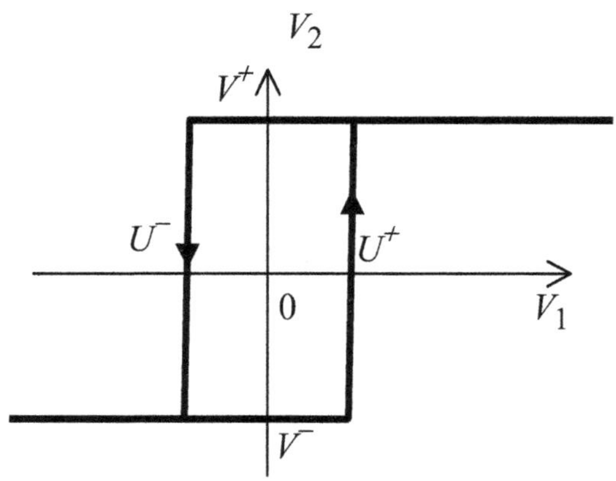

ii) If K is in position (1) $V_C(t) = \frac{I_0 t}{C} + V_{C1}$ and in position (2) $V_C(t) = -\frac{I_0 t}{C} + V_{C2}$.

If $V_1 = V_C$ increases, corresponding to position (1) of K, V_2 will become equal to V^+ corresponding to condition (a); therefore if (a) $\Rightarrow$ K = 1, the system will be latched at $V_2 = V^+$ and will remain at this value. To obtain an oscillation, the reverse should apply, namely:

$$V_2 = V^+ \Rightarrow \text{K}=2,\ V_2 = V^- \Rightarrow \text{K} = 1$$

In this case, $V_2 = V^+$ leads to $V_C(t) = -\frac{I_0 t}{C} + V_{C1}$, therefore decreasing, which allows $V_1 = V_C$ to become smaller than U^-; this will lead to flip-flopping $V_2 = V^-$. Then, $V_C(t)$ is increasing, which leads to V_C becoming higher than U^+ and so on. The operating point describes a rectangular cycle, as shown in below figure ($V_2(V_1)$).

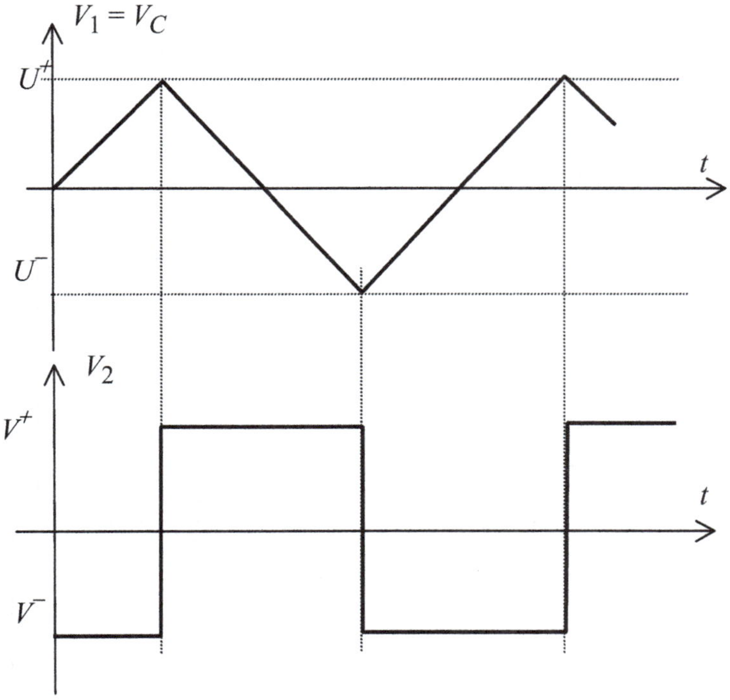

The positive slope of $V_C(t)$ being $\frac{I_0}{C}$, the half period is $2\frac{C}{I_0}\frac{R_1}{R_2}V^+$ supposing $V^+ = -V^-$. This leads to frequency: $\frac{I_0}{4CV^+}\frac{R_2}{R_1}$ which becomes

$\dfrac{V_e}{4CR_0V^+}\dfrac{R_2}{R_1}$ if I_0 is a source of current controlled by voltage V_e (or I_0 = V_e/R_0): $f_2 = \dfrac{1}{4CR_0V^+}\dfrac{R_2}{R_1}V_e$.

4.6.3. *Noise figure of an operational amplifier circuit*

i) Making use of the circuit indicated in the course for noise sources at operational amplifier input, draw the full diagram of a non-inverter circuit featuring feedback network R_1 (between minus input and ground) and R_2 (between minus input and output). The source of signal to be amplified has an internal resistance R_g. Find the noise figure of the amplifier circuit as a function of elements, sources of noise and bandwidth Δf.

ii) Calculate the conditions under which the noise figure is minimum, supposing that R_g is variable and can be adjusted.

Answer:

i)

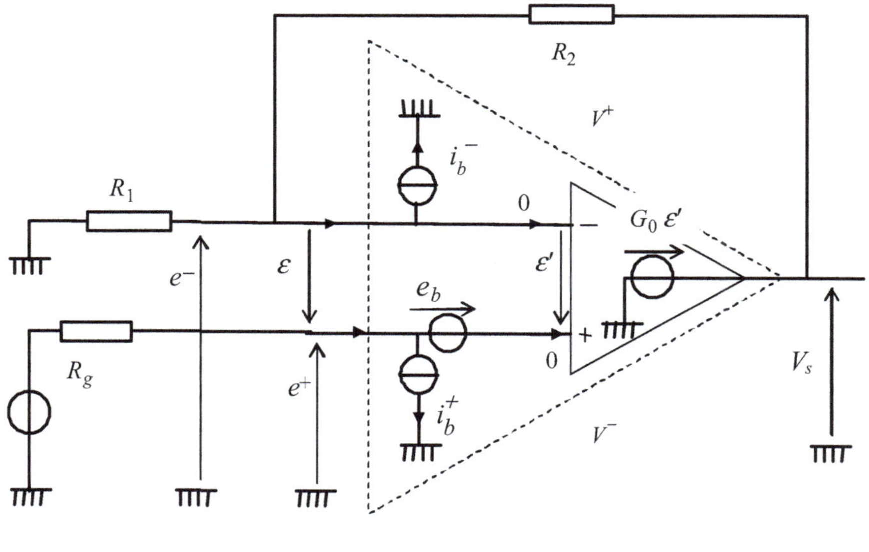

To calculate the noise figure, the theorem of superposition is used, determining first the contribution of each source of noise. On the + input:

– for resistance R_g, there is a source of noise voltage $\sqrt{4kT\,R_g\,\Delta f}$ in series;

– noise current i_b^+ generates a noise voltage $R_g\,i_b^+$ across this resistance;

– noise source e_b is in series with the two previous ones.

On the – input, the contribution of noise current i_b^- distributed through resistances R_1 and R_2 and considered as alone is negligible if the operational amplifier voltage gain is far above 1, which is the case in the bandwidth. Indeed, since in the calculation the other independent sources are considered cancelled, voltage is null on + input. On – input, we then have the sum of three currents: $\varepsilon/R_1 + (G_0 + 1)\varepsilon/R_2 + i_b^- = 0$; this leads to the rms value (sign is not considered):

$$\varepsilon = \frac{i_b^-}{\dfrac{1}{R_1} + \dfrac{G_0 + 1}{R_2}}$$, which tends to 0 if G_0 tends to infinity.

Finally, the effect in rms value of the three sources present on the – input is added, which yields the total mean squared noise voltage: $e_{btot}^2 = 4kT\;R_g\,\Delta f + (R_g i_b^+)^2 + e_b^2$.

The noise figure is obtained from the definition:

$$F = \frac{e_{btot}^2}{4kT\,R_g\,\Delta f} = 1 + \frac{(R_g i_b^+)^2 + e_b^2}{4kT\,R_g\,\Delta f}$$

ii) If R_g is considered variable, F can be differentiated in order to determine the existence of an extremum:

$$\frac{dF}{dR_g} = \frac{8kT\,R_g^2\,\Delta f\,(i_b^+)^2 - \left((R_g i_b^+)^2 + e_b^2\right)4kT\,\Delta f}{\left(4kT\,R_g\,\Delta f\right)^2} = 0$$

which is null if $(R_g i_b^+)^2 - e_b^2 = 0$ and therefore: $R_g = \frac{e_b}{i_b^+}$. This is a minimum, since F tends to infinity if R_g tends to 0 or infinity.

Therefore, there is an optimum of the noise figure for this value of R_g and also for an operating frequency within the frequency band $[f_{c1}, f_{c2}]$, where the two bounds are the "corner" frequencies of spectral densities of voltage noise and current noise, respectively. This optimization is important for the amplification of signals issued by low-level sensors and for low-noise RF amplifiers.

Appendix

Electrical Circuits

A.1. Laws of electrokinetics for linear passive elements

Elements	Circuit and conventions	Instantaneous expressions	Complex symbolic regime†	Causal regime*
Resistor (Resistance R)	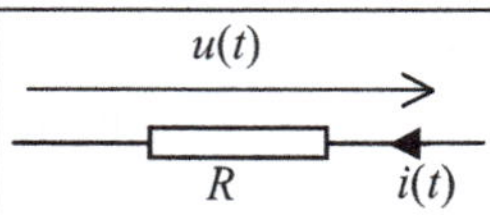	$u(t) = R\,i(t)$ (Ohm's law)	$U = R\,I$	$U = R\,I$
Capacitor (capacitance C)	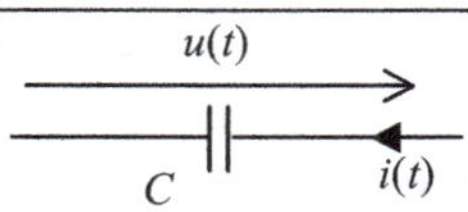	$i(t) = C\,\dfrac{du(t)}{dt}$	$I = jC\omega U$	$I = C\,s\,U$
Inductor (Inductance L)	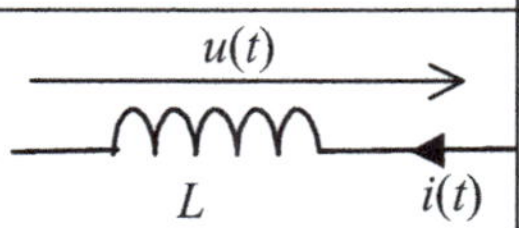	$u(t) = L\,\dfrac{di(t)}{dt}$	$U = jL\omega I$	$U = L\,s\,I$

†For sinusoidal, periodical regimes or for signals with existing Fourier transform. Please see substantiation at the end of this appendix.
*Obtained by Laplace transform of voltages and currents if their values before time origin are null (forced regime) for capacitor and inductance.

Considering the signs of currents and voltages such that currents flow from higher potentials (head of the voltage arrow) to lower potentials (origin of the voltage arrow), corresponding to a receptor convention, the laws are written with a plus sign for passive elements. If the direction of current or voltage is reversed, each law is written with a minus sign.

Each coefficient R, C or L corresponds to a relation of proportionality, more commonly called linear law. Nevertheless, this linearity is not a general property and passive elements should rather be defined based on quantities related to electrokinetics, electrostatics and electromagnetism as in the following section.

If elements are placed in series, and consequently the same current I flows through them, the sum of terms that are factors of I is an impedance, noted Z in symbolic or causal regime. If elements are placed in parallel, and consequently submitted to the same voltage U, the sum of terms that are factors of U is an admittance, noted Y in symbolic or causal regime.

A.2. Definition of passive elements

More generally, passive elements follow nonlinear current–voltage relations (resistance of a junction, inductance of an electric coil with magnetic core, etc.). They can be defined by a derivative around a specific value (bias point) of related electrical quantities such as in the table below (V is the voltage or electric potential difference, I is the intensity of electric current, Q is electric charge, Φ_B is the magnetic flux and Φ_E is the electric flux as defined below). In a first approximation, the value thus defined is maintained, as well as an affine relation between the two electrical quantities, as long as the deviation from the effective value remains within a predetermined range. Otherwise, the value of the element should be modeled by means of a nonlinear function (or limited development), which is easily carried out when using a simulator.

Type	Definition	Units
Resistance	$R(I) = dV/dI$	V/A or Ω
Capacitance	$C(Q) = dQ/dV$	Coulomb/V or F
Inductance	$L(I) = d\Phi_B/dI$	Weber/A or H
Memristance	$M(Q) = d\Phi_E/dQ$	(V.s)/Coulomb or Ω

Electric flux Φ_E is defined as the integral of voltage across an element with respect to time (V·s being its units), which confers memory properties to memristance, a nonlinear nature and a link with the magnetic flux Φ_B through Lenz law in an inductance, as $d\Phi_B/dt$ has voltage dimension. Since memristance is equal to the derivative of Φ_E with respect to the charge crossing this element (memristor), ohm can be considered its unit of measurement, as equivalent to voltage to current ratio, and itself a quotient of charge to time. This element is subject to debate, research, patent registration, controversies and clarifications since the 1970s, both on theoretical and experimental levels. It will not be considered in this work.

A.3. Ideal sources

A.3.1. *Voltage sources*

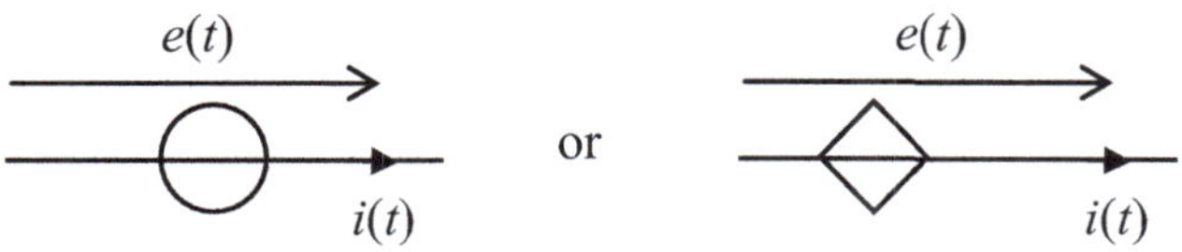

(sign conventions for generator)

The most frequently used symbol is a circle crossed by a line, the lozenge being sometimes used for controlled sources, whose voltage (or electromotive force) is proportional to a voltage or current in another circuit branch.

In direct current (DC) with an electromotive force E_0, the characteristic is plotted below.

The current in a voltage source depends on the value of the electromotive force e <u>and</u> of the external circuit. Therefore, it is not uniquely determined by the source.

If $E_0 = 0$, this is a <u>null voltage source, equivalent to short circuit.</u>

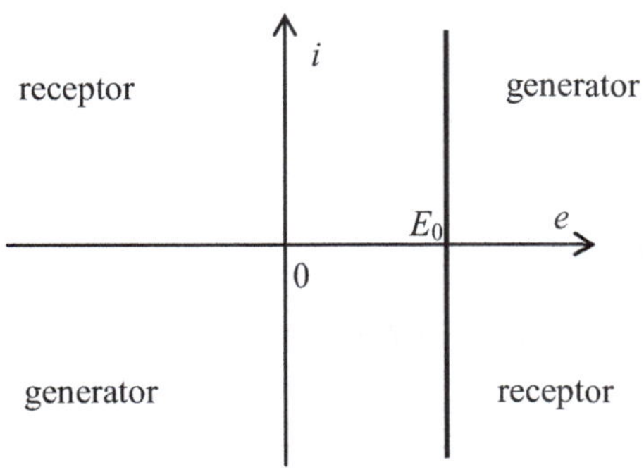

A.3.2. *Current sources*

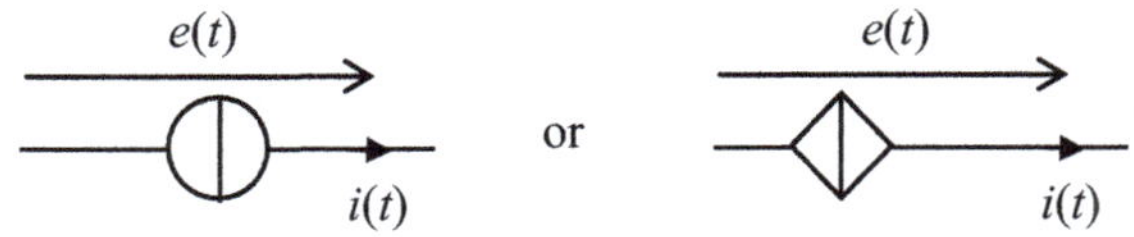

(sign conventions for generator)

The most frequently used symbol is the circle crossed by a line perpendicular to the conductor line, the barred lozenge being sometimes used for controlled sources, meaning sources whose current is proportional to a voltage or current in another circuit branch.

In direct current, with a current I_0, the characteristic is plotted below. The current in a voltage source depends on the value of the current i and on the external circuit. Therefore, it is not uniquely determined by the source.

If $I_0 = 0$, it is a null current source, equivalent to an open circuit.

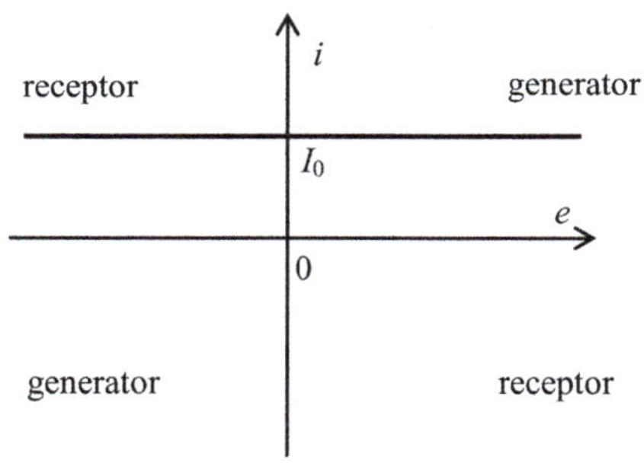

An ideal source that is not controlled by another one is independent.

In regimes other than the direct one, notions of generator and receptor depend on the average active power (please refer to Volume 2 [MUR 17a]).

A.4. Conservation laws

A.4.1. *Voltage law in close and open loops*

The potential difference along a closed loop is null, like the corresponding vector sum in the next figure, with Z_k being impedances:

$$Z_1 I_1 + Z_2 I_2 - E_1 - Z_3 I_3 - V_4 = 0 \text{ (algebraic or complex notations);}$$

or still the potential difference between two points of a circuit is the same, whatever the path followed, or:

$$V_0 = Z_1 I_1 - V_4 = Z_3 I_3 + E_1 - Z_2 I_2$$

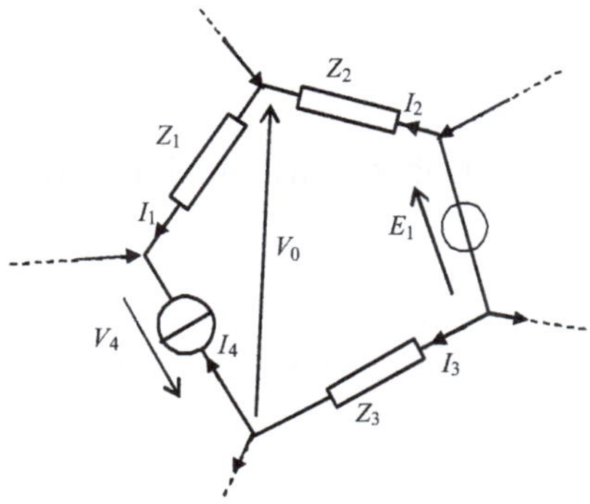

A.4.2. *Current law at a node*

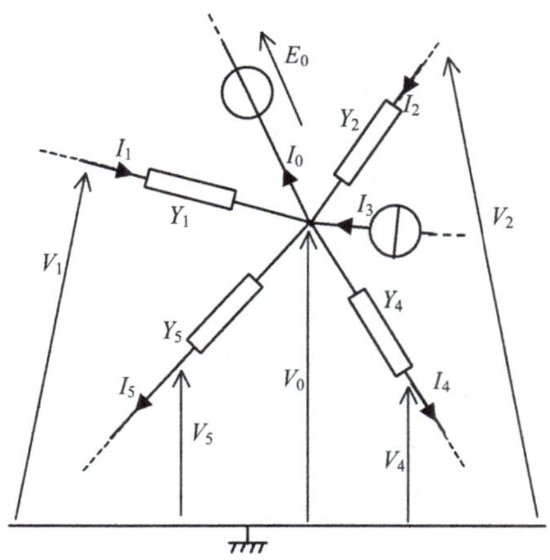

The sum of currents (algebraic or complex notations) in a node is null:

$$I_1 - I_0 + I_2 + I_3 - I_4 - I_5 = 0$$

or as a function of potential differences in the previous figure, with Y_k being admittances:

$$(V_1 - V_0)Y_1 - I_0 + (V_2 - V_0)Y_2 + I_3 - (V_4 - V_0)Y_4 - (V_5 - V_0)Y_5 = 0$$

This is because no charge can spontaneously appear or disappear in a conductor.

An electric network comprises loops and nodes through which sources and passive elements are electrically connected. If it has no connection with another network, it is isolated.

A.5. Number of state variables and independent equations in an isolated network

In a network, state variables are currents and voltages or potential differences. There are as many currents as there are branches, namely b. Since there is a voltage difference across two nodes, and the number of nodes is n, there are $n - 1$ voltage differences across nodes. The current flowing through a branch and the voltage at its terminals are interdependent due to electrokinetics laws, except for branches featuring an ideal source, in which one of the two quantities depends on the external circuit. This last type of branch should not be taken into account.

There are therefore $b - n + 1$ independent state variables. Determining all these values requires a system of linear equations with unique solution, therefore with the same number of independent equations, $b - n + 1$. This number is also the rank of the system of linear equations.

In the common case of finding a relation between two quantities, as in electronics when finding input and output gains, resistances or impedances in a circuit, the rank of the system should be decreased by one unity (system called "undetermined" in mathematics, as it has as many solutions as values given to one of the two quantities), which is obtained with $b - n$ equations for the same number of state variables.

A.6. Useful theorems

These theorems result from the linearity of previous laws. In a completed circuit, the previous laws give rise either to a system of linear equations in which currents are the unknowns and the sources form the second members of these equations:

$\mathbf{Z}\,\mathbf{I} = \mathbf{E}$ in a matrix form

or to a system of linear equations in which voltages are the unknowns and independent sources form the second members of these equations:

$\mathbf{Y}\,\mathbf{U} = \mathbf{I}$ in a matrix form

Controlled sources appear in the first members as they are proportional to the quantities that play the role of unknowns, therefore in the first member. The solution is obtained by inverting matrices:

$\mathbf{I} = \mathbf{Z}^{-1}\,\mathbf{E}$ or $\mathbf{U} = \mathbf{Y}^{-1}\,\mathbf{I}$

Taking the previous solutions into account, the equation related to the branch containing impedance Z_1 can be written in the form $V_1 = E_{int} - Z_1\, I_1$ or $I_1 = I_0 - Y_1\, V_1$.

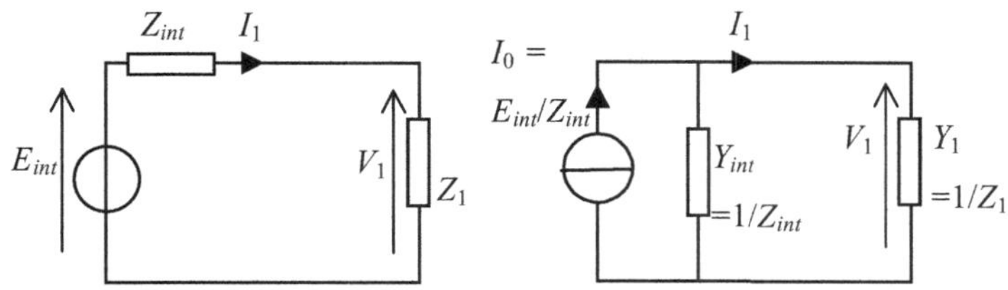

Any circuit seen between two points can be replaced either by a voltage generator composed of an ideal voltage source and an internal impedance (Thévenin's theorem) or by a current generator composed of an ideal source of current and an internal admittance (Norton's theorem). The voltage in open circuit is E_1 in two cases and the short-circuit current is $E_1/Z_1 = E_1 Y_1$ in two cases. The two circuits are fully equivalent.

Impedances Z_{int} and Z_1 form a voltage divider and voltage V_1 can be deduced with the relation $V_1 = \frac{Z_1}{Z_1 + Z_{int}} E_{int}$; admittances Y_{int} and Y_1 form a current divider and current I_1 can be deduced with $I_1 = \frac{Y_1}{Y_1 + Y_{int}} I_0$.

Millman's theorem relates to the parallel association of sources. In the following circuit, after transforming voltage generators into current generators, the law at central node gives $\frac{E_2}{Z_2}+\frac{E_3}{Z_3}=Y_2E_2+Y_3E_3=(Y_1+Y_2+Y_3)V_1$, which leads to $V_1=\frac{Y_2E_2+Y_3E_3}{Y_1+Y_2+Y_3}$.

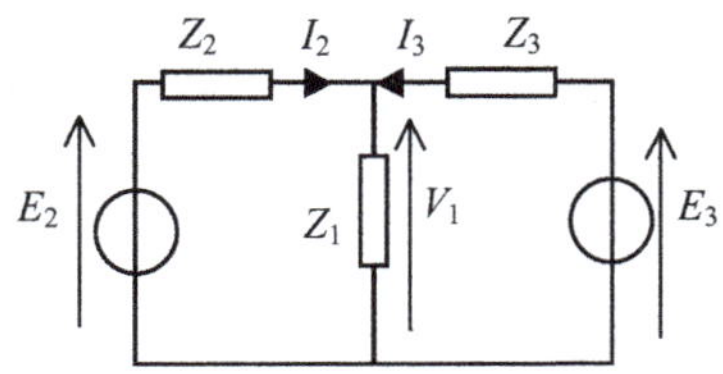

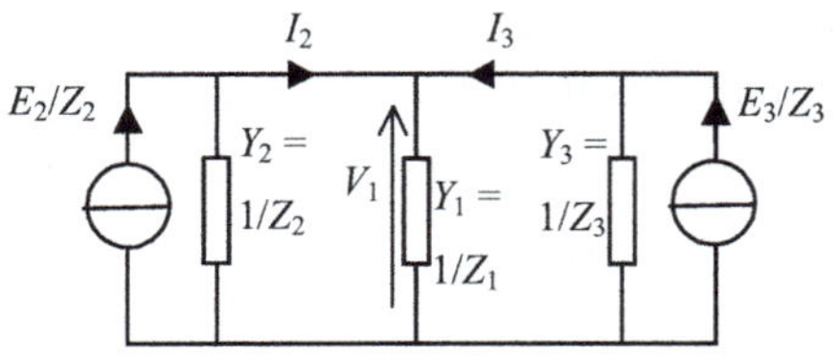

Superposition theorem can be deduced from matrix equations:

$$\mathbf{I}=\mathbf{Z}^{-1}\,\mathbf{E} \text{ or } \mathbf{U}=\mathbf{Y}^{-1}\,\mathbf{I}$$

in which column matrices containing independent sources **E** or **I** have been decomposed into a sum of matrices. For example:

$$E=\begin{bmatrix}E_1\\E_2\\E_3\\E_4\\E_5\end{bmatrix}=\begin{bmatrix}E_1\\0\\0\\0\\0\end{bmatrix}+\begin{bmatrix}0\\E_2\\0\\0\\0\end{bmatrix}+\begin{bmatrix}0\\0\\E_3\\0\\0\end{bmatrix}+\begin{bmatrix}0\\0\\0\\E_4\\0\end{bmatrix}+\begin{bmatrix}0\\0\\0\\0\\E_5\end{bmatrix}$$

in each of which only one element is non-zero, corresponding to only one non-null source. In one or the other of the first members, the result is obviously the sum of products of $\mathbf{Z}^{-1}$ (or $\mathbf{Y}^{-1}$) by each matrix. The superposition theorem can therefore be stated: any current or voltage in a circuit branch results from the summation of effects of each of the ideal sources present in the circuit and considered successively, the other being replaced by null sources (short circuit for independent voltage sources and open circuit for independent current sources).

A.7. Circuits in non-steady state

In a non-steady state linear circuit, voltages and currents in elements such as inductance and capacitance have to be considered in terms of their instantaneous values. Solving the circuit equations then leads to an ordinary differential equation of pth order in which $x(t)$ and $y(t)$ represent each one of the instantaneous electrical quantities of the network and where a_k and b_k depend on passive elements:

$$\sum_{k=0}^{p} b_k \frac{d^k y(t)}{dt^k} = \sum_{k=0}^{m} a_k \frac{d^k x(t)}{dt^k}$$

because $\exp(\alpha t)$, where α is real or complex, is an eigen function of this differential equation, since if $x(t) = A\ e^{\alpha t}$ and $y(t) = B\ e^{\alpha t}$, the differential equation is reduced to a characteristic polynomial equation $B\sum_{k=0}^{p} b_k\ \alpha^k = A\sum_{k=0}^{m} a_k\ \alpha^k$, without variable t. This suggests the systematic use of complex exponentials as functions representing electrical quantities.

Moreover, for the whole network studied, the number of independent quantities should be counted based on the number of differential equations of first order that can be written by introducing the number of intermediate quantities necessary and sufficient for differentiating these quantities only once. Another approach is then possible by solving the differential system of first order (see Volume 2 [MUR 17a]).

A.8. Example of sinusoidal regime in an RLC circuit

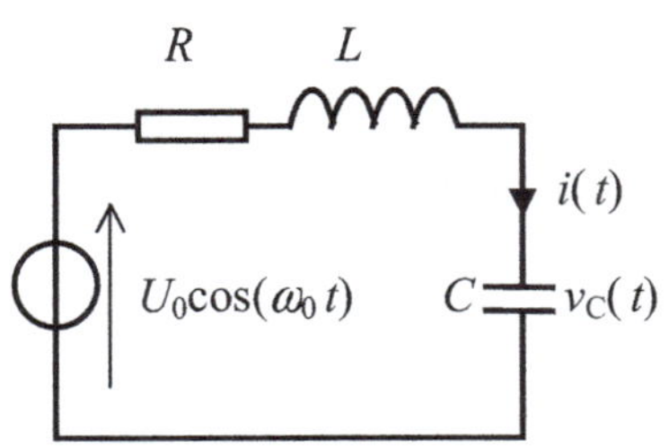

$$U_0 \cos(\omega_0 t) = Ri(t) + L\frac{di(t)}{dt} + v_C(t) \text{ and } i(t) = C\frac{dv_C(t)}{dt}$$

There are two quantities and two equations that involve only first-order derivatives. The number of independent variables is therefore 2. Elimination of $i(t)$ between these two equations leads to:

$$LC\frac{d^2v_C(t)}{dt^2} + RC\frac{dv_C(t)}{dt} + v_C(t) = U_0\cos(\omega_0 t)$$

Replacing $U_0 \cos(\omega_0 t)$ with:

$$\frac{U_0}{2}\left[\exp\left(j\omega_0\, t\right) + \exp\left(-j\omega_0\, t\right)\right]$$

and considering

$$v_C(t) = V_{C0}\cos(\omega_0 t+\varphi)$$

or still

$$v_C(t) = \frac{V_{C0}}{2}\left[\exp\left(j\omega_0\, t+\varphi\right) + \exp\left(-j\omega_0\, t+\varphi\right)\right] = \frac{V_C}{2}\exp\left(j\omega_0\, t\right) + \frac{\overline{V_C}}{2}\exp\left(-j\omega_0\, t\right)$$

where V_C is complex, the differential equation can be written in the following form:

$$\begin{aligned}&U_0\left[\exp\left(j\omega_0 t\right)+\exp\left(-j\omega_0 t\right)\right]\\&=V_C\left[LC(j\omega_0)^2+jRC\omega_0)+1\right]\exp\left(j\omega_0 t\right)\\&+\overline{V_C}\left[LC(j\omega_0)^2-jRC\omega_0+1\right]\exp\left(-j\omega_0 t\right)\end{aligned}$$

This equation cannot be verified for all the values of t unless after grouping together the coefficients of $\exp(j\omega_0 t)$ and $\exp(-j\omega_0 t)$, the factors of two exponentials are null:

$$U_0 = V_C\left[LC(j\omega_0)^2 + jRC\omega_0 + 1\right] \text{ or } V_{C0}e^{j\phi} = \frac{U_0}{LC(j\omega_0)^2 + jRC\omega_0 + 1}$$

The factor of $\exp(-j\omega_0 t)$ yields exactly the complex conjugate of the result obtained when the factor of $\exp(j\omega_0 t)$ is null, therefore it is not necessary to write it.

To reach this result, voltages can be considered complex $V_C = V_{C0}\, e^{j\varphi}$, U_0, as well as the current $I_0\, e^{j\varphi'}$. In fact, if equality $U_0 = V_C\left[LC(j\omega_0)^2 + jRC\omega_0 + 1\right]$ is rewritten with $jC\omega_0$ in factor, we obtain:

$$jC\omega_0 V_C \left[jL\omega_0 + R + \frac{1}{jC\omega_0} \right] = U_0$$

which represents the product of current evaluated from the product of complex voltage V_C and complex admittance of the capacitor $jC\omega_0$ multiplied by the sum of complex impedances of inductance, resistance and capacitance, respectively. The three elements being in series, this yields the total voltage across the source of voltage U_0. Quantities such as V_C and U_0 are commonly called symbolic notations.

A.9. Conclusion

The previous calculation is valid for angular frequency ω_0, also called the circular frequency. It corresponds to representation of electrical quantities in a regime in which voltage and current variables are complex exponential functions, factor of $\exp(j\omega_0 t)$, which disappears after simplification. This is called complex symbolic regime, as it does not represent all the sinusoidal signals, since complex conjugated quantities corresponding to operation at angular frequency $-\omega_0$ should also be added, in order to find the real quantities, which are representative for the physical reality. But since the complex conjugate part satisfies exactly the same relations after complex conjugation, there is no need to consider them. The use of complex representation, which remains valid irrespective of signal frequency in circuits with localized elements, therefore also for multifrequency signals such as periodic signals decomposable into Fourier series or having a Fourier transform, considerably simplifies circuit analysis. Due to the linearity of laws valid for linear circuits and to the validity of the previously mentioned superposition theorem, it is indeed sufficient to superpose as many complex symbolic representations as there are various frequencies in the signals.

Bibliography

[BLO 93] BLOT J., *Electronique linéaire*, Dunod Université, Paris, 1993.

[GIR 90] GIRARD M. (ed.), *Composants Actifs Discrets* (Volumes 1 and 2), McGraw Hill, London, 1990.

[HOR 02] HOROWITZ P., HILL W., *The Art of Electronics*, 2nd ed., Cambridge University Press, Cambridge, 2002.

[MUR 17a] MURET P., *Fundamentals of Electronics 2*, ISTE Ltd, London and John Wiley & Sons, New York, 2017.

[MUR 17b] MURET P., *Fundamentals of Electronics 3*, ISTE Ltd, London and John Wiley & Sons, New York, 2017.

[SHA 97] SHAEFFER D.K., LEE T.H., "A 1.5-V, 1.5-GHz CMOS low noise amplifier", *IEEE Journal of Solid-State Circuits*, vol. 32, no. 5, pp. 745–759, 1997.

[SZE 07] SZE S.M., NG K.K., *Physics of Semiconductor Devices*, 3rd ed., John Wiley and Sons, New York, 2007.

Index

T, V, Z

Other titles from

ISTE

in

Electronics Engineering

2017

BUCCI Davide
Analog Electronics for Measuring Systems

2016

BAUDRAND Henri, TITAOUINE Mohammed, RAVEU Nathalie
The Wave Concept in Electromagnetism and Circuits: Theory and Applications

FANET Hervé
Ultra Low Power Electronics and Adiabatic Solutions

NDJOUNTCHE Tertulien
Digital Electronics 1: Combinational Logic Circuits
Digital Electronics 2: Sequential and Arithmetic Logic Circuits
Digital Electronics 3:Finite-state Machines

2015

DURAFFOURG Laurent, ARCAMONE Julien
Nanoelectromechanical Systems

2014

APPRIOU Alain
Uncertainty Theories and Multisensor Data Fusion

CONSONNI Vincent, FEUILLET Guy
Wide Band Gap Semiconductor Nanowires 1: Low-Dimensionality Effects and Growth

Wide Band Gap Semiconductor Nanowires 2: Heterostructures and Optoelectronic Devices

GAUTIER Jean-Luc
Design of Microwave Active Devices

LACAZE Pierre Camille, LACROIX Jean-Christophe
Non-volatile Memories

TEMPLIER François
OLED Microdisplays: Technology and Applications

THOMAS Jean-Hugh, YAAKOUBI Nourdin
New Sensors and Processing Chain

2013

COSTA François, GAUTIER Cyrille, LABOURE Eric, REVOL Bertrand
Electromagnetic Compatibility in Power Electronics

KORDON Fabrice, HUGUES Jérôme, CANALS Agusti, DOHET Alain
Embedded Systems: Analysis and Modeling with SysML, UML and AADL

LE TIEC Yannick
Chemistry in Microelectronics

2012

BECHERRAWY Tamer
Electromagnetism: Maxwell Equations, Wave Propagation and Emission

LALAUZE René
Chemical Sensors and Biosensors

LE MENN Marc
Instrumentation and Metrology in Oceanography

SAGUET Pierre
Numerical Analysis in Electromagnetics: The TLM Method

2011

ALGANI Catherine, RUMELHARD Christian, BILLABERT Anne-Laure
Microwaves Photonic Links: Components and Circuits

BAUDRANT Annie
Silicon Technologies: Ion Implantation and Thermal Treatment

DEFAY Emmanuel
Integration of Ferroelectric and Piezoelectric Thin Films: Concepts and Applications for Microsystems

DEFAY Emmanuel
Ferroelectric Dielectrics Integrated on Silicon

BESNIER Philippe, DÉMOULIN Bernard
Electromagnetic Reverberation Chambers

LANDIS Stefan
Nano-lithography

2010

LANDIS Stefan
Lithography

PIETTE Bernard
VHF / UHF Filters and Multicouplers

2009

DE SALVO Barbara
Silicon Non-volatile Memories / Paths of Innovation

DECOSTER Didier, HARARI Joseph
Optoelectronic Sensors

FABRY Pierre, FOULETIER Jacques
Chemical and Biological Microsensors / Applications in Fluid Media

GAUTIER Jacques
Physics and Operation of Silicon Devices in Integrated Circuits

MOLITON André
Solid-State Physics for Electronics

PERRET Robert
Power Electronics Semiconductor Devices

SAGUET Pierre
Passive RF Integrated Circuits

2008

CHARRUAU Stéphane
Electromagnetism and Interconnections

2007

RIPKA Pavel, TIPEK Alois
Modern Sensors Handbook

Printed and bound by CPI Group (UK) Ltd, Croydon, CR0 4YY
13/09/2026
14961040-0001